The Business of Construction Contracting

The Business of Construction Contracting

Schleifer's Guide to Financial Success

Thomas C. Schleifer, Ph.D.
Aaron B. Cohen, MS, CPC

Library of Congress Cataloging-in-Publication Data has been applied for:

Print ISBN   9781394279111
ePdf ISBN   9781394279135
ePub ISBN   9781394279128
oBook ISBN   9781394279142

Cover Design: Wiley
Cover Image: © CoreDESIGN/Shutterstock

Set in 9.5/12.5pt STIXTwoText by Lumina Datamatics

Contents

Preface

This book provides the tools, strategies, and insights needed to build a sustainable and profitable construction business. Whether you are an aspiring entrepreneur, an established contractor looking to refine your business practices, a construction professional seeking to understand the broader industry dynamics, or a student of construction, this book will serve as your guide.

The construction industry is one of the oldest and most essential industries in the world. It shapes the built environment that surrounds us. The ability to successfully complete projects is at the core of what makes a constructor, but it does not inherently equip someone with the skills required to manage a thriving construction business. Many construction professionals enter the field with technical expertise and hands-on experience to produce the work but soon realize that success in business also requires a completely different skill set – one that is often overlooked in trade, engineering, and construction education.

This book bridges that gap. It provides a comprehensive guide to running a construction business efficiently and profitably, equipping construction professionals with the necessary knowledge to navigate the complexities of the modern construction industry. Readers will learn how to identify the common elements of construction business failure, how to structure a business, and how to select the right projects and effectively manage the business throughout the ever-changing landscape of the construction industry.

A key theme explored in this book is the challenge of growth. The very traits that make constructors successful in the early stages of a business, such as hands-on problem-solving, direct involvement in project decisions, and a relentless work ethic, can become obstacles to sustainable expansion. As a company grows, so does the need for structured management, delegation of authority, and strategic planning. Construction professionals must transition from working *in* the business to working *on* the business. Without this shift, many managers fail to understand that growth and the rate of growth include the risk of financial instability and business failure.

The complexity of the construction industry has increased significantly over the years. Regulatory requirements, supply chain disruptions, labor shortages, and heightened client expectations demand greater management abilities than ever before. Even small construction businesses must now navigate intricate legal and financial issues, risk mitigation strategies, and workforce management challenges. Without a firm grasp of these elements, even the most skilled builders can find themselves struggling to keep their companies afloat.

Compounding these challenges is a notable generational shift in the industry. Fewer young professionals today aspire to own a construction business, preferring instead the stability of employment over the uncertainties of entrepreneurship. At the same time, mergers and acquisitions are reshaping the competitive landscape, with small firms increasingly being absorbed by medium-sized regional firms that are also increasingly being absorbed by larger and multinational corporations. As the industry consolidates, independent small and midsize construction firms must adapt to new ways of competing and differentiating themselves in an evolving market.

All these factors contribute to one crucial reality: success in construction contracting today is no longer just about delivering a quality project – it is about running a business effectively. The ability to manage cash flow, hire and retain talent, navigate regulations, leverage technology, and adapt to industry changes is what separates thriving construction companies from those that struggle to survive.

The business of construction contracting is evolving rapidly, and those who embrace change, invest in their business expertise and adopt a forward-thinking approach will be best positioned for long-term success. Let's get started.

About the Authors

Thomas C. Schleifer, Ph.D., joined the construction industry at age 16 and has more than 60 years of contracting and consulting experience. He has Bachelor of Science and Master of Science degrees in construction management from East Carolina University and a Ph.D. in construction management from Heriot-Watt University in Edinburgh, Scotland. Dr. Schleifer's experience includes serving as foreman, field superintendent, project manager, and vice president of a construction company that he owned with his brother. From 1976 to 1986 he was the founder and president of the largest international consultancy firm serving the contract surety industry. During this period, he assisted in the resolution or salvage of hundreds of distressed or failed construction firms.

This combination of practical, hands-on experience as a contractor and assisting financially distressed companies has given Dr. Schleifer a unique perspective on the causes of construction business failures and how to avoid them. Dr. Schleifer, sometimes referred to as a "turnaround" expert because of the number of companies that he has rescued from financial distress, advises contractors on organization, structure, and strategic planning while he also writes, lectures and teaches.

The importance of education in the construction industry is one of Tom Schleifer's favorite themes. He has lectured extensively at universities and professional and trade associations and authored numerous articles on construction and business management. Dr. Schleifer has been listed in *Who's Who in Finance and Industry*, *Who's Who in America*, and *Who's Who in the World*. He was the 1993 Eminent Scholar of the Del E. Webb School of Construction, Arizona State University.

Publications by Dr. Schleifer include books *The Business of Construction Contracting*, 2025, John Wiley and Sons; *The Secrets To Construction Business Success*, 2022, Routledge Taylor & Francis Group; *Managing the Profitable Construction Business*, 2014, John Wiley and Sons; *Construction Contractors' Survival Guide*, 1990, John Wiley and Sons; *Glossary of Suretyship and Related Terms*, CMA and weekly a blog *Schleifer's Weekly Construction Message*.

Aaron B. Cohen, MS, CPC, is the Director of Estimate Products at InEight Inc., where he defines product requirements and oversees the development of software solutions for the construction industry. Prior to joining InEight, Aaron held the Associated General Contractors (AGC) Lecturer position at Arizona State University in the Del E. Webb School of Construction, where he continues to teach courses in Infrastructure Estimating and Project Controls.

Aaron also spent more than 10 years as the president of a trenchless engineering and construction services provider specializing in the application of trenchless technologies for public construction projects. He has extensive experience managing and estimating infrastructure and utility construction projects and has lectured on the subjects of Horizontal Directional Drilling and underground utility construction for various industry associations, including the American Society of Civil Engineers (ASCE), American Water Works Association (AWWA), American Public Works Association (APWA), Underground Construction Technology (UCT) Expo, and the North American Society for Trenchless Technology (NASTT) No-Dig show, and has been an instructor at the Horizontal Directional Drilling Academy since its inception in 2014.

Aaron received a Bachelor of Science degree from Arizona State University as well as a Master of Science degree from DePaul University and is a Certified Professional Constructor. He is the author of *The Business of Construction Contracting*, 2025, John Wiley and Sons; and has co-authored the books *Construction Planning, Equipment, and Methods, 10th edition*, 2024, McGraw-Hill; *Moving the Earth, 7th edition*, 2019, McGraw-Hill; and has made contributions to the *Handbook for Building Construction*, 2021, McGraw-Hill; and *Horizontal Directional Drilling (HDD) Good Practices Guidelines, 4th edition*, 2017, NASTT.

1

The Construction Industry

Construction may be the oldest industry in the world, but it was not until 1990 that it was discovered to have the second-highest failure rate of any industry in the United States. This finding was the result of 12 years of research that also documented the specific causes of construction business failure. And perhaps more significantly, the causes of failure were determined to be preventable. The research process included a detailed analysis of hundreds of annual financial statements prepared by independent accounting firms. A "statistically significant number" of all types and sizes of failed construction companies were analyzed in detail. The results were compared and contrasted with an equal number of similar non-failed companies over the same time period. It took years of research to identify the primary causes and in some cases the secondary causes of the failures. Most accounting experts were shocked to discover that the underperformance and losses discovered were occurring two to three years prior to losses being reported on the construction companies' certified financial statements.

After the publication of this research, a lot of construction companies altered their behavior and prospered. However, a large number of contractors stated that while the findings were correct, the lessons did not apply to themselves but pertained to less informed and less skilled companies that failed. The rationale, of course, was: *We have not failed so the findings don't pertain to us.* Unfortunately, hundreds who held on to that belief subsequently did fail. Many of those who failed said to us or wrote to us with comments like: *I heard you but didn't believe it* or *I told others to change their behavior but didn't change my own* or *I didn't want people to think I didn't know what I was doing.*

Few dispute that construction is a high-risk endeavor, and our research demonstrates that being in the construction business for a long period of time offers little protection. It may sound counterintuitive, but one of the difficulties in convincing construction professionals that these common elements of business failure are accurate and high risk is that the elements *don't* always cause business failure. For example, there is a huge risk in taking a project that is much larger than the organization has ever built or taking a project in an unfamiliar geographic area. However, when a firm takes either of these risks and it works out successfully, they believe they have disproven the research, and that the size of the project and change in geographic area is not a common element of construction business failure. These firms then, of course, continue to embrace the risk in the future because they did not believe the risk exists.

The Business of Construction Contracting: Schleifer's Guide to Financial Success, First Edition. Thomas C. Schleifer and Aaron B. Cohen.
© 2025 John Wiley & Sons, Inc. Published 2025 by John Wiley & Sons, Inc.

The reality is that while the scientifically proven common elements of construction business failure do not always cause failure they are, nevertheless, hugely risky because they often cause failure, just not 100% of the time. Many make the mistake of trying to measure the risk by measuring the odds. They think something like 6 to 1 odds are better than 2 to 1. The fallacy of measuring the odds this way is the true measurement is not simply the odds that an occurrence might happen. The real measurement must include the odds of an occurrence AND the magnitude of the occurrence in question. For example, most reasonable people would readily accept a bet of $1 at 6 to 1 odds. The question is would they still be interested at 6 to 1 odds if the bet was a million dollars?

Six to one odds with a bet of $1 is inviting for two reasons; one it would be great to win $6 and two the loss of $1 is worth it because it doesn't really hurt. The loss of a million dollars makes the bet seem almost ridiculous at any odds. An analogy I sometimes use is Russian roulette. If you take a six-shot revolver and place only one bullet in it, spin the cylinder, point it at your head, and pull the trigger; is it a small risk or huge risk? You might like 6 to 1 odds at a bet of $1, but few would bet their life at 6 to1 odds.

If a larger project or a project out of your normal geographic area has the potential to weaken your financial position to the point of eventual failure, is it a reasonable business risk or a huge (almost ridiculous) risk? Indisputable scientific research data demonstrate that these activities have put a "statistically significant" number of other contractors out of business. Even if you are convinced you can beat the odds because you have been in business a long time, do you really want to "*bet the farm*"? If you were an investment advisor, would you be recommending these risks to your clients? You may also find it of interest to know how many of the failed construction firms in the database were second- and third-generation companies demonstrating *that length of time in business* does not protect against business failure.

The information about construction business failures we have to draw on today was expanded with the addition of 30 more years of research data. While the causes of business failure have modified some over time and the industry has grown tremendously in size, construction still remains the second-highest failure rate in the nation (*second only to restaurants*). Many of the construction professionals that were exposed to the original data have retired by now, and most of the new generation of industry leadership were never exposed to these essential documented research findings. The authors believe it is important to update and restate this material for the new generation of construction leadership. Fortunately, the industry has progressed, and leadership has benefited from specialized engineering and construction education programs developed over the last several decades, advancing the construction process considerably.

On the downside, very few of these education programs address the business side of the business of construction. Therefore, we are able to build better, but we are still having the second-highest failure rate. The industry failure rate continues, including spectacular failures like the 2022 multinational Carillion organization, the largest construction enterprise in the world. In light of the current US construction boom and global economic uncertainties, the failure rate will undoubtedly continue and has the potential to substantially increase. This information is most critical to a more receptive industry than ever.

The industry's growth combined with the persistent nationwide labor shortage has increased contractor risk and will, in all likelihood, increase the industry failure rate. Similar historic conditions in the past resulted in construction business failures, which makes the solutions in this text that much more significant. We need to understand how constructors can succeed during various and changing market conditions. This is particularly significant because most construction professionals and managers are graduates of engineering or construction schools or came up through the ranks as tradespeople. Engineering or construction education curriculum teaches how to capture and produce construction work, but typically includes little or no business, accounting, economics, finance, statistics, marketing, or personnel management courses.

This text provides a shortcut into how to structure and manage the business side of the business of construction with guidance on:

- Proven construction business strategies
- The advantages of and how to develop short- and long-term business plans
- The concepts of flexible overhead
- Corporate, and financial self-analysis
- The element of construction contractor failure
- An understanding of construction market cycles
- Cash flow thresholds
- The threat of industry accounting weaknesses
- The true cost of equipment ownership
- Modern leadership techniques
- The latest industry technologies and client maintenance
- Global construction industry trends
- And more

This book includes links to proprietary software programs, including:

- Company self-analysis tool
- Project selection tool
- Corporate self-analysis tool
- The R-score calculator
- Schleifer's manual of construction practice

Construction businesses underperform or fail not because they cannot produce the work, but primarily because they are not managed as a profitable "business" concern.

1.1 A Brief History

The modern construction industry in the United States as we know it today commenced at the end of World War II (WWII) when soldiers returned from the war and received benefits from the Servicemen's Readjustment Act of 1944, also known as the GI Bill, providing accessible and affordable mortgages. A building boom began fueled by housing demands that created unprecedented employment opportunities (Figure 1.1). Since then, in spite of temporary recessions, the construction industry has continued to grow through today. The industry prospered through the 70s paying wages well above typical factory work and generating double-digit profits. Not long after that and almost unnoticeable the construction wage advantages over other industries began to slip, and profits slowly drifted along with them. About the same time (in reaction to this or by coincidence, it is difficult to tell), new contracting methods began to be introduced, altering the "competitive balance" of the old-line competitive process.

Figure 1.1 Post–World War II building boom fueled by housing demand created a prosperous time for contractors.

Construction Management (CM), heretofore unheard of, became the latest procurement method and some contractors believed it was great because they thought it would reduce their project risks while maintaining profit margins. Neither occurred as CM fees became miniscule compared to prior profit margins and, unbeknownst to contractors, they were still saddled with project completion in accordance with plans and specifications risks. The procurement method mutated into various permutations of CM, such as CM at Risk, Guaranteed Max, and various other limitations.

Then came Design-Build contracts, which designers originally jumped into as lead partners until they rapidly tired of the risks they did not realize they were undertaking. Then contractors jumped in with both feet without realizing they were *"in effect"* guaranteeing the design quality and function. As new procurement methods multiplied, court actions were bound to follow. It may be hard to believe, but it looks like low bid is making a comeback. It is beginning to appear that the only thing procurement experimentation accomplished was to drive down profit percentages across the board, which will be permanent unless the construction industry does something about it.

For both newcomers to the industry and old-timers, this history is hard to believe because it happened so slowly and because few top managers today have ever worked in the industry when contractors were regularly making two-digit profits. A less obvious, but very real ramification of the declining margins over time has created a less attractive industry. One of the more obvious signs is fewer family businesses. Our children see how difficult the industry is, the long hours we work, and the low margins we make excuses for. Most are too well-informed to follow in our footsteps. As a result, the industry is consolidating, which means that big contractors are getting bigger, small contractors stay small while midsize contractors (with no willing heirs) consolidate, liquidate, or sell out. The number of midsize contractors will be drastically reduced because there is no way they can accommodate the overhead necessary to maintain themselves or make enough profit to be worth the effort. They can't effectively compete against large contractors who survive on greater volume at drastically reduced profits with a much smaller percentage of overhead costs. The economy of scale is very real, particularly under current market conditions. The next question has to be:

1.2 What's Ahead?

The current state of the construction industry works well for the buyers of construction services (owners) because they take the position that they should not and will not share any risks as a result of the construction process. The designers aspire to the same position. Having considerable control over the construction contract documents (i.e. AIA contract documents, and the like), they continue to modify the contracts to provide the designer with total "authority" over the design while accepting no "*responsibility*" for the design. Contractors regularly work under contract language that states that the contractor must notify the designer if a design element will not work as intended, and if the contractor proceeds with the work, they are responsible for it. These difficult contract provisions will outlive us unless we do something about them. In the meantime, while we must follow the contract language specifically, we can also use the contract language to protect ourselves. This is addressed in the subsequent chapters.

The industry will continue to grow into the foreseeable future because as the country grows in population and wealth, the need and want for bigger and better shelter and facilities expands. There have always been cyclical downturns and recessions, but over the 80 years since WWII, the country and national wealth have grown continually in spite of short-term reversals. The built environment is an integral and large part of our nation's Gross Domestic Product (Figure 1.2).

Contrary to industry belief there has always been more than enough work to go around, which calls into question another industry belief. That getting enough work is the main function of top management. This is inappropriate, in error, and a leading driver of the industry's failure rate. The principal function of construction company leadership and all construction professionals for that matter, is to earn a profit from safely and efficiently producing the work (the built environment). This, and not continuous growth, is the only long-term sustainable business goal and is the guiding principle of this text. Construction industry participants sometimes need to be reminded that "profit is not a dirty word."

Figure 1.2   The built environment is an integral and large part of our nation's Gross Domestic Product.

1.3   Industry Beliefs

Construction being perhaps the oldest industry in the world happens to bring a lot of old beliefs with it. It is fair to say that the longer we are in this business, the more "baggage" we bring along with us. Beliefs however are not always factual and some beliefs that may have been true in the past may not be true today.

An expert on beliefs, author and philosopher Barry Kaufman says that:

Everyone, without exception, holds to beliefs that are largely unexamined, but are often our deepest conviction, and that we constantly collect evidence to support them. He goes on to say: It is almost impossible to dislodge these beliefs, even if they are patently false and threaten our well-being in the future.

This book tests a number of construction industry beliefs, the first of which is that expanding or growing a construction business is always good and should be a prime objective of any or every construction organization. In fact, many construction professionals will tell you "*If you are not growing, you are going backward.*" Having spent a large part of our careers researching the causes of construction business failures, the authors can assure you this is patently false and can be extremely dangerous. We have asked any number of failed construction company leaders why they went after a currently distressed and losing project that was so far outside of their organization's collective experience. Inevitably the answer was: "*I was running out of work.*" Other answers that were popular were: "*If the work is there go after it; We'll figure out how to do it later; We're not afraid of risk; We can build anything.*"

Most of these long-held beliefs sound good and seem like they actually may make good sense. Questioning these strongly held industry beliefs is not easy. And it is compounded by the fact that contractors who have succeeded in this ridiculously risky and difficult business have a lot to be proud of. Add to this that none of us care to have our hard work and strongly held beliefs questioned, and it is an uphill battle to say the least. These are critical topics, so let's take it one step at a time. As for the above comments about the need for work or running out of work, these are addressed later in this book and hopefully to your satisfaction. For now, let's explore some of the realities of the construction marketplace that impact our need for profitability and actually constrain our ability to capture work at a profit.

This may be a good place for all of us to agree that if you bid cheap enough you can get all the work you want. This is just a reminder that the primary objective of being in business, other than a *charity or not-for-profit business*, is to make a profit. At the risk of offending anyone, if the profit motivation does not apply to you, you probably don't want to read this book. For those still interested let's explore the mathematical certainty that the construction marketplace is the prime factor affecting the potential for profitable work.

1.4 An Example

We have watched this principle dawn on construction professionals many times while assisting company planning groups attempt to compile their first-time multi-year business plan. Prior to the planning session, we would have distributed industry growth projections for the next several years for their segment of the industry in the geographic area they work. Early in the planning session, the group would be charged with coming up with sales expectations and goals for the next several years that they can all agree on. They would be encouraged to take into consideration their desired goals, the company's resources, and available work. Inevitably, the groups would be discussing growth projections of 10, 15, 20%, or more for the next year and the years after. Shortly into these discussions, the group would be asked if they considered the projections made by the state chamber of commerce, their local contractors' association, national contractors' associations, and any other published forecasts we may have given them. They would also have been given the average of all these data, which they would be reminded of (for example) was 7% growth for each of the next two years. The typical response is *yes, but that does not place any controls on us because our goals are separate and distinct from the forecasts.*

While forecasts do not place any controls on planning groups and their projections may turn out to be higher or lower, they definitely impact profit potential going forward. Whether the industry forecasts are correct is not the point. They are all you have. Let's consider what we do know. If our sample market grows at 7% and all practitioners grow at 7%; mathematics and logic suggest that the prior "competitive balance" should not alter much. This is because bid amounts and profits are under little or no pressure to reduce and may be able to increase.

If the same market grows at 7% and the company in question (and perhaps others) intend to capture enough work to grow 15 or 20%, their increase MUST result in and be at the expense of other practitioners in the same market because the others will capture proportionately less work. Mathematically, other businesses must get smaller for any business in the same market to grow at a larger percentage than the overall market grew. This is a fact, because the overall market has a finite size, such as 7% growth. Another historic reality is that to take that work away from other bidders who also want that work you have to "bid aggressively," which means lowering your price.

This should be obvious to construction professionals, but it is not necessarily the way we think about our market because most of us have little motivation to subject this topic to a cause-and-effect analysis. We are fully entitled to want to grow faster than the market, but it defies logic to think that will occur if we stick to our old pricing because our competitors will be offering their old pricing or better depending on their growth plans. Some readers may be tempted to say fine we have always competed so what's new. Nothing is new. If the pie is a certain size and we want more than our historical mathematical share, others will have to take less or fight back by lowering their prices. It's that simple. Competitive balance is a concept successful contractors of the future will fully understand and once they do, it will impact and inform their pricing. More than one contractor has been heard to say: *I know how to do this, just don't change the rules on me*. We are not changing the rules. Just shedding some light on the rules that are less obvious.

1.5   How Many Ways?

Another belief we need to address early on is that: *We contractors can run our business anyway we darn well please*. This may, in the literal sense, be true, but the statement is incomplete from a business standpoint. It should actually be: *We contractors can run our business for a profit anyway we darn well please*. If we all followed the second statement, our industry would not have the second-highest failure rate in the nation. There are well over a million construction contractors in the United States, and while there is more than one way to run a construction business for a profit, there are not anywhere near a million ways. There may be half a dozen ways and with minor differences maybe a dozen but that is it. In our work with failed construction organizations, we have experienced hundreds of different ways to run a construction enterprise, which is part of the reason our industry has such an abysmal failure rate. When an organization is managed in a random or informal manner, it inevitably achieves random inconsistent results.

To be clear, we are addressing here the business side of the business of construction as opposed to the technical or performance of the work side of the business. With plumbing, plastering, or painting; building, heavy/highway, or tunneling; general contractor, subcontractor, or construction manager, the technical side or performance of the work is different, specialized, and unique to each trade. However, the business side of every construction enterprise is not just similar, they are exactly the same (Figure 1.3).

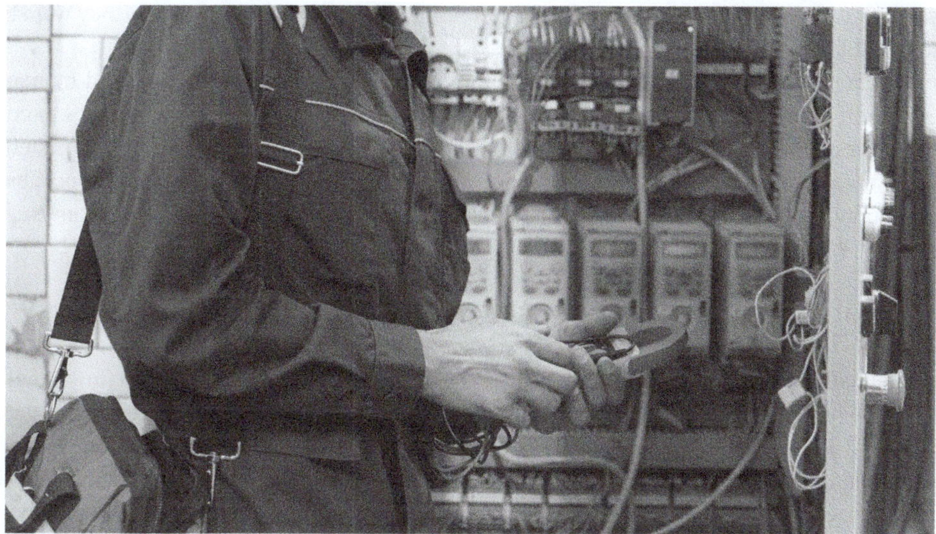

Figure 1.3 While the technical skills employed by each trade are unique and specialized, the business of construction is identical for all trades.

This will be widely disagreed with by many because just about every trade argues that their business is different from the others. They are correct on the technical side of the enterprise, but they are not on the business side of the enterprise. Taking this a step further, if any construction professional is running their businesses differently from the best business practices outlined here, they could be doing better and they may be doing it wrong.

For better or worse, how to structure and manage the modern construction organization was historically developed by trial and error. Those doing it well enjoyed varying degrees of success. Those doing it poorly resulted in withdrawal or failure. Both good and poor organizational structure or informal personal-preference management and accounting processes have existed for years. The good processes succeeded. The poor processes failed, or in a few cases underperformed for years until their luck ran out. The question of course is, how do we know which construction business management, marketing, and accounting practices are best? Quite simple, results.

In this complex industry, there is an interesting disparity in historic performance as measured in profitability. In most industries, the worst and best profit performers are separated by small percentages. In construction, the worst and best profit performance is measured by factors of two or four times and, in some cases, more. This makes little sense when you consider that, in theory, each of the firms had access to the same resources, drew from a similar labor pool, and worked in similar markets under the same conditions. Our research demonstrates that the differentiator is the caliber and quality of business practices. This is confirmed by our experience in the construction company turnaround business.

In turning around hundreds of underperforming or failing construction enterprises, the only changes we made were within the business side of the enterprise in question. In all cases, if the underperformance was within the production side of the business, turnaround was almost impossible. The reason was obvious.

If a construction organization could not properly and profitably perform their specialty, they were by definition not a viable enterprise.

Few researchers have seen firsthand and had more access to the financial details and management practices of both failing and world-class performers. We were privileged to conduct studies to compare and contrast these processes, practices, and data, isolating best practices that generated consistent profits. These best practices have been dissected and deconstructed so that they can be understood, absorbed, and put into practice by readers who want to advance to world-class construction professionals.

1.6 Evaluating a Construction Organization

The authors break down the business side of the business of construction into three functional areas: getting the work, doing the work, and accounting for the work. These can also be expressed as marketing, production, and administration. This may seem an oversimplification, but it identifies the essence of a construction organization and highlights the reality that three separate and distinct talents and skill sets are critical for success. As these are clearly separate categories of education, profession, and experience, few, if any, individual businesspersons would be trained and skilled in each area.

As noted earlier, few engineering or construction curriculum have room for many business courses and few business, marketing, or accounting curriculum have room for construction or engineering courses. While most contractors, particularly small and midsize consider themselves a *"jack of all trades,"* few would agree with the old saying that describes this as *"but master of none."* Our experience is that most people, contractors included, may be experts in one of these three fields that they were educated and experienced in, but not in two or three.

A simple case in point. When my brother and I formed our construction business, both of us were builders, and neither of us were marketers or accountants. We got by *as so many do*, but becoming world class is a struggle at best. Obviously, experts can be hired, but most contractors are not aware of exactly what skills are required. It is possible that few readers will be aware until now that these three skill sets are equally critical to the success of any construction enterprise. This will surely be debated; however, the fact that numerous construction companies failed because of weakness in only one of the three areas should put that debate to rest. Some failed enterprises were weak in two areas and a few in three, but many failed that were expert in two of the areas and weak in only one.

These three are equal in significance (even critical), not because they contribute equally to corporate profitability, but because each can cause business failure. While earning a profit is generally considered primary in any business. In construction, protecting the enterprise is a close second and some would argue co-primary with profit.

So, what does this mean to contractors and senior construction professionals with leadership responsibilities? One important message is to leverage your expertise with a full understanding of its limits and recognizing that none of us knows everything. An exceptional builder that came up through the ranks as an engineer would have had limited time for much formal education in accounting, marketing, or business management studies. A person's success at building the work (production), combined with their intelligence, does not mean that you are now an expert in getting the work (marketing) or accounting for the work (finance and accounting). These are separate and distinct professions that take years of education and experience to master. These experts can and are hired, but the problem is that the contractor with no proficiency in the field in question is not, in fact, qualified to supervise the positions.

This is a far more serious issue than our industry recognizes in that mistakes are made, and companies fail when the experts are ignored, disagreed with, or countermanded. Because the majority of contractors are the producers of the work, the critical business areas of "getting the work" and "accounting for the work" are very regularly reduced to "advisory" capacities. Critical information is passed along to a contractor (owner or boss) with little or no education or experience in the discipline – who then makes decisions on what advice to accept or reject. It may seem harsh to say, but by definition, the owner or boss is not qualified to make those decisions. Some will say, that's the way it is, so what choice do we have. The first choice is to recognize and accept that this is a dysfunctional management structure as verified by industry results – the second-highest industry failure rate. This is further verified by decades of research into the causes of construction business failure.

The solution is not complicated. Start by recognizing that the three functional areas of a construction enterprise are equal. Not because they contribute equally to profitability but because missteps in any one of them can and has, singlehandedly, caused numerous construction firms to fail. While it is important to concentrate on doing things to maximize profit, it is critical to avoid things that might even remotely risk failure. Once this is understood and accepted, the next step is to equalize the input from the leadership of the three functional areas of getting, doing, and accounting for the work.

During the day-to-day management of a construction firm, the persons in the roles of production, accounting, and marketing are critical to the success of the organization. Obviously, if the firm does not capture enough or appropriate work, it suffers. If the firm can't produce the work at a profit, it suffers. And if the firm does not accurately and timely account for the work, it suffers. There may be differences between the amounts of profit each role contributes to the firm, but they are meaningless in the context of survival. Mistakes or mishaps in any of these areas can be costly, including the potential for catastrophic failure. Many profitable construction operations were triggered by the loss of banking or bonding relationships because of losses or because they could not pay their bills. A construction firm can lose either of these relationships because of unfavorable accounting reports or because it turns in no financial reports.

The three functional areas are equal, critical, and must work closely together to the point of being integrated closely. The authors have data on companies that failed because they did not capture enough work, the right type of work, or work in the right location. Others failed because they produced a large enough portion of their work at a loss sufficient to create a losing year or two in a row. And still others failed because they lost control of their accounting data and could not produce accurate annual financial statements, in a timely manner, and in a number of cases at all.

We can't count the number of people who have argued that producing the work is all that matters in a construction operation. While it is true that for most enterprises, production of the work is the major contributor to profits. It is also the major contributor to failures, so some might say that balances out. The authors keep saying that prevention of failure is equally important as generating a profit, and the more it is discussed, the more important it seems to become.

The remainder of the book treats generating profit and preventing loss as equal, and readers will need an open mind because if our industry is going to improve our position from the second-highest failure rate in the nation, we are going to have to change some of our beliefs and some of the ways we do business. Change is always difficult.

2

Structure of a Construction Business

Construction is unlike other major US industries in terms of the number of companies in the industry. Other heavy industries, such as oil, mining, aviation, and shipping, have perhaps dozens of large, leader firms and hundreds of smaller companies. The majority of the large companies are publicly traded, and, as such, their performance information is widely known.

The construction industry is composed of over a million businesses. These businesses range from one- or several-person operations to the very largest companies. All but a dozen or so of these firms are closely held, family-owned, or have few owners and, as part of a very horizontal industry, tend to not share detailed performance information. With no clear industry leaders, unlike in other industries, construction companies have limited guidance on the ideal business structure, operation standards, and performance metrics.

This situation is further complicated by the differences in the sizes of companies that are structured and operate very differently. We categorize the companies as small, medium, large, and jumbo according to the following definitions:

- Small: The owner(s)/contractor(s) personally perform some or all of the work and manage the business.
- Medium: The owner(s)/contractor(s) do not personally perform the work but rather direct and supervise the people responsible for the work and regularly visit project sites.
- Large: The firm has departments and/or divisions that are responsible for performing the work, and the owner(s)/contractor(s) have limited, if any, contact with the work. They seldom visit any work sites and may be informed about field activities primarily through written reports and perhaps occasionally at meetings.
- Jumbo: The firm is typically organized into divisions according to geographic area or type of work. Each division operates similarly to how a large firm operates – usually with separate organization charts, chains of command, and so forth. Basically, the divisions look like and perform as separate large construction companies that report to a common owner.

The Business of Construction Contracting: Schleifer's Guide to Financial Success, First Edition. Thomas C. Schleifer and Aaron B. Cohen.
© 2025 John Wiley & Sons, Inc. Published 2025 by John Wiley & Sons, Inc.

2.1 Organizational Structure

A common weakness in the construction industry is that many firms are structured ineffectively: they have unrecognized chains of command, they don't have departments or divisions, and they are loosely organized. One of the reasons for this ineffective structuring of the business is that many, probably most, of these companies were begun by one person, who learned how to do the work by coming up through the trades or through technical or engineering education. These individuals lack business or management education and believe that all that matters is producing the work. Thus, because they started out small, they didn't think a formal organizational structure was needed. Working hard and hiring similarly minded people was sufficient. And as start-up companies experience success early, the idea that a formal organizational structure is not necessary to be successful gets reinforced and becomes a belief of the contractor.

Drawing on research, we determined that many more start-up construction companies fail within 10 years than continue operating. Because the defunct companies were small and unstructured, there is no way to verify whether having a formal structure may have helped. Consequently, the research we conducted on the common element of construction business failure does not include any companies in business for fewer than 10 years. From examining a statistically significant number of failed and operating construction companies, we determined that construction companies typically failed during the first generation; some companies lasted into the second and third generations. In most of the first-generation companies, an inappropriate management structure was determined to be a factor in the failure. The research on the structures of construction businesses began with these firms.

The structure of an organization should be built around the three primary functions of a construction enterprise, which were introduced in Chapter 1: getting the work (marketing/sales), doing the work (production/operations), and accounting for the work (finance/administration). The many other functions, such as personnel, equipment management, purchasing, etc., all fall under one of the three functional areas. We found that failed companies almost universally lacked structure or had inappropriate structures and that successful companies had efficient and appropriate structures. It is that simple. Construction professionals who decide to invent their own corporate structure should ask themselves if inventing business structures is what they're experts in.

Some will argue, "We don't need an organizational chart because everyone knows what their positions are and what they're supposed to do." When we consulted with successful companies, the majority of managers we interviewed would say things such as "I know exactly which direction I'm supposed to be moving in because the boss talks about it all the time." The next manager we talked to would say the same thing but would refer to the opposite direction.

Our experience is that the construction workforce is willing and qualified to do the work. The industry's main weaknesses are organization and leadership. It bears repeating; the preference for an informal structure is a weakness. Most owners and supervisors tell us they don't want a lot of rules and procedures. They want their employees to have the freedom to do their best. In contrast, employees say that all they need is direction. If someone would clarify what they're supposed to do, they'd be doing it. Instructing people to just do their best doesn't work anywhere. Suppose we told soldiers on a battlefield, "there's no rank or orders. Everyone do what you think is best."

Direction, as well as order and a clear, concise chain of command, is the basics of efficiency. Each employee either works for someone or works for no one. The argument that a company is too small for

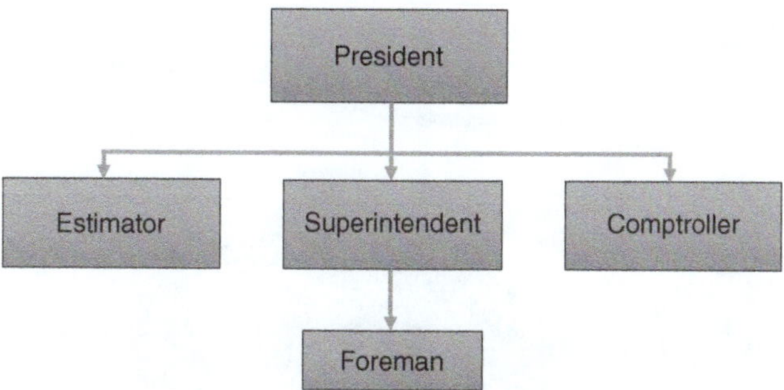

Figure 2.1 Sample organizational chart for a small construction company.

an organizational chart doesn't hold water. The answer to the question "What size do we need to be to have a formal organizational chart?" is: "From day one." The reason is that once you get going, it's too difficult to introduce order where there has been none. Wherever we've introduced a formal structure, often referred to as a *reorganization*; productivity and job satisfaction have improved immediately and dramatically. Organizational charts (Figure 2.1) are easy to create, and how to create one isn't nearly as important as why to. You can follow the examples shown in the chapter, or you can refer to the multitude of other examples available. Often, first-timers will realize that some employees currently have no one to answer to. A gap in the chain of command is a message to employees to solve the situation on their own. In successful businesses, everyone answers to someone. Even the chairman of the board answers to the board.

2.2 Quality Decision-making

With an organizational structure in place, the next steps in operating the business are to decide what to do, who will do it, and how. Before we address these questions, let's revisit the construction industry's typical chain of command. Historically, the contractor – the boss – has called all the shots. However, it's obviously time to revisit that structure, considering that the industry has the second-highest failure rate in the nation. We need to ask ourselves, if one person can run a business, would the potential for success improve with added brainpower and experience? If so, would two or more managers collaborating be more effective?

We contend that most experienced businesspeople and logical thinkers answer yes to both questions. The next obvious question is, should the added experience be similar to the boss's experience or be different? Clearly, someone with similar skills and background may be easy to hire and may feel like a more comfortable addition, but such an individual would bring only limited additional benefit. Therefore, different but complementary experience is preferred. In the construction industry, when experts in the three functional areas of the business confer on major decisions, these experts will clearly produce results that are superior to what one person could achieve (Figure 2.2). The only objection to this perspective that we can envision is that this process may include short delays. Research on decision-making suggests that increasing the speed of decision-making reduces the quality of the decisions and is appropriate only in emergency situations.

Figure 2.2 Employing managers with different but complementary experience results in better decision-making.

Let's look at an example: A contractor took a call from his head estimator while waiting in his car at a red light. The estimator said there was a project out for bid that was larger than any they had ever done and asked whether he should even consider it. The contractor said, "Pick up the plans and see what it looks like." Several days later, the contractor got into the office to look at the plans. The contractor said, "Well, it's not really our type of work, but you've already put some effort into it, so let's put in a bid." We learned about this situation when we were called in by the contractor's bonding company long after the contractor got the job and lost a huge amount on it, approaching bankruptcy. Slowing down the decision-making process and openly discussing the project with the senior managers of all functional areas of the organization may have prevented this disaster. Slowing down the decision-making process for high-level questions always improves the quality of the conclusions or it verifies initial inclinations before committing to them.

Why did the company fail on this project decision? We learned after the fact that when the head estimator called the contractor, he was already convinced they should not attempt this project. However, the contractor ran such a tight ship that no one could make a decision of this nature without clearing it with him. The estimator had also learned from experience that this contractor was too busy to listen to advice from others and tolerated no pushback once he had announced a decision. This busy contractor had inadvertently sown the seeds of his own destruction. There's a good chance that the company wouldn't have failed if the contractor had cared about the opinions of his senior managers, or at least listened to them. Collaborating makes even more sense.

Was one busy person who happened to also own the company the best person to make the decision? The contractor would say, "I had to because that's my responsibility." That statement is true, but it was his responsibility alone only because he made it his responsibility alone. Being the sole owner doesn't prevent anyone from forming an executive committee and empowering the group to make senior-management decisions. The biggest roadblock here is self-imposed. Success, particularly entrepreneurial success, has a tendency to cause people to believe that they are chosen or special. Some begin to believe that they are orders of magnitude smarter than others, particularly their employees. (After all, if employees had the talent or brainpower, they would be running their own businesses, right?) It's easy for any of us to forget that experts are almost always experts in one thing. On the other hand, there are people who know a lot about a lot of things. Just ask them – they will tell you.

The reason we're spending this much time on this topic is that few people in our industry realize how large this problem is, and even more don't want to hear that it is even a problem. We need to understand that experts in construction nationwide are supervising, directing, and making decisions for their accounting and marketing functions without having passed the Certified Public Accountants (CPA) exams or taken any of the marketing, advertising, and selling courses, their vice president (VP) of marketing has.

It may surprise some, but who owns a business is meaningless in measuring the qualifications of senior managers and should never be considered when measuring an executive's level of expertise. A marketing executive cannot judge the quality of accounting or a construction decision any more than a constructor can judge the quality of accounting data or marketing efforts. Nevertheless, in our industry, the prevailing belief is that whoever owns the company decides everything. Most contractors tell us that they do make all the decisions, but only after considering or evaluating the input from their Chief Financial Officer and marketing VP. What most do not seem to recognize is that all the construction experience in the world does not qualify a construction expert executive to evaluate the information provided by experts in fields unrelated to construction like accounting or marketing. If you employed a nuclear physicist, would you therefore understand fission or fusion? Ownership does not trump expertise, nor does it create or inspire it.

2.3 Collaboration

The best way to achieve secure, profitable business results is to engage in a collaborative effort that benefits from the expertise of the leaders of the three primary functional areas of the business and any others who are experts in the subject at hand. Being the owner of the company offers no added value in the decision-making process about the management of the business. In decision-making, the value of the input from each functional area is equal even though the profit each functional area contributes is not equal. The differences in profit contributions are known to even modestly informed decision-makers and have no bearing on the intellectual process of decision-making.

The collaborative process works only if the value of the information from each decision-maker is considered equal. If the input from the functional area that regularly contributes the greatest portion of profit is given greater weight in any manner, then that functional area might as well just make all the decisions. We have come full circle. That approach doesn't work and is the problem we're attempting to solve.

If you're wondering why we're emphasizing this issue, it's because the industry has decided that the business owner is entitled to make all the decisions. This book isn't a thesis on business ownership or about entitlement. This book is a thesis on business management and business success. Specifically, construction business management and success. We are addressing an uncomfortable reality. The construction industry has accepted an inappropriate belief about the decision-making process, which contributes to the industry having the second-highest failure rate in the nation.

In discussion groups on this topic, contractors usually agree with our position but still maintain that as the business owners and the ones at risk, they should unilaterally make the final decisions in certain categories. However, there are no categories relating to the management of the business in which any one discipline should be the sole decision-maker. Some have suggested that all major business decisions should be made at the board of director level. We respectfully disagree.

Contractors tell us, *"If I disagree with my employees' positions, they're usually wrong. I hired them, they work for me, and it's my money, so why should I listen to them?"* These statements may be true, but as we stated earlier, the construction expert isn't qualified to judge accounting information unless the construction expert has passed the CPA exam. Unfortunately, many contractors disagree with us.

2.4 A Superior Process

We fully realize that the collaboration we recommend will require a significant shift from current processes. However, most things remain the same. For example, collaboration in decision-making doesn't alter the chain of command. The boss is still the boss and remains responsible for the success of the organization. Collaboration, as defined here, dramatically improves the quality of decision-making and doesn't imply that the person ultimately responsible doesn't remain so. The owner is responsible for announcing unanimous decisions and for making decisions after considering all of the input when decisions aren't unanimous. Of course, there will be disagreements when collaborating, requiring compromise and tiebreaking. This fact doesn't negate the fact that collaboration hugely improves the quality of decision-making.

Further, we aren't implying that collaboration will always result in the best decisions possible, will always be correct, and that it will cause decision-making to be easy. What collaboration does is put an end to decision-making that doesn't consider all the information and opinions available. It slows the process, minimizing *spur-of-the-moment* errors, and as a side benefit energizes non-owner managers to maximum effectiveness because they feel more empowered to do their job to the best of their ability. When their input is respected and valued, and they see how their involvement can be influential to the objectives of the company, people tend to outperform themselves.

Collaboration can be facilitated through many processes using whatever strategy works for your company. Some people prefer regularly scheduled meetings, some prefer as-needed meetings, and yet others favor various other options. It should go without saying that face-to-face meetings are best, but they're not always possible. What doesn't work is separate meetings with different decision-making parties because each party must hear what everyone else says. They also need to observe how they say it so they can read between the lines. What absolutely doesn't work is deciding that whoever knows most about the topic will make the decision and clear it with the others. Collaborative decisions require that a group discuss the issues and work together on the solution.

One of the difficulties with instituting true collaboration in decision-making is the potential of discovering that one or more managers aren't really qualified for the positions they have. This discovery is actually positive in that it is a group to then weed out weak links. Such a discovery is an inconvenience and may cause embarrassment, particularly if the weakness involves family members or long-time employees. This is a situation that is one of the difficulties unique to family and closely held companies, common in the construction industry. Identifying weak links happens often enough that you should be prepared for it, embrace the improvements that eventually result, and move on. This process, along with identifying people who report to no one, results in an improved organizational structure on the way to world-class performance. World-class construction companies do not have ineffective or inappropriate business structures.

2.5 Business Planning

It's appropriate to provide a reminder at this point that this book isn't about how to construct the work but rather about how to manage a construction business profitably, which includes recognizing, managing, and mitigating risks. We never minimize how critical it is to produce the work, because no amount of business skills can save an organization that is not capable of prosecuting the work at a profit. Failure to profit is, by definition, not a *business* because businesses are *for-profit* organizations.

This book is intended for construction enterprises that can produce the work but aren't structured and organized in a manner that maximizes profitability and minimizes risk. The data presented about the failure rate in construction were collected from "successful" construction enterprises that failed. All companies in the database were successful and in business for 10 years or more. No start-up companies were included because the startup failure rate in all industries is huge; therefore, including start-ups would skew the data.

With this reminder out of the way, let's discuss the next step to take after a construction enterprise is appropriately structured and has a functional chain of command. The value of long-term planning and strong business strategies cannot be overstated. Throughout history, people who have succeeded in business have planned. From ancient hieroglyphics, we know that builders made plans for the great structures still standing today. All contractors plan but tend to spend most of their time and effort planning the work. Many seem to lose sight of the absolute necessity of planning for the business side of the enterprise.

A construction professional would not attempt to construct a building, bridge, or pipeline without a design, schedule, workforce, and materials. As familiar as contractors are with the necessity and value of good planning in terms of construction projects, you'd expect the contractors would be the first to embrace and engage in comprehensive business planning to manage their enterprises.

In an informal study, fewer than half of the companies claimed to do any type of corporate planning, and very few of these companies said they regularly referred to their business plans after they were developed. The most common reason given for not planning was the organization's size. Believing that small and midsize construction businesses don't need to plan is a serious business error. Small companies with limited resources can't maximize their potential with a trial-and-error approach. Of the million-plus construction companies in the United States, the largest percentage are small businesses. In this tough and competitive industry, too many small contractors believe they can succeed by personally and aggressively driving their businesses forward. This approach seldom works, and the few times that it does, it is definitely not an easy way to succeed.

We can't overstate the importance of comprehensive business planning regardless of a company's size. Likewise, we can't overstate the impact that the lack of planning has on the likelihood of business failure. Contractors have underestimated business planning for too long, and if this trend can't be reversed, the industry's failure rate will remain the same or get worse. The prosperity of the industry depends on the ability to make some dramatic changes, not the least of which is realizing that our granddad's old tried-and-true ways of running a business were for the old days and are way past their use-by date.

We need to put an end to the industry's painful weeding-out process, in which companies that don't keep up fail. Some people believe that this process is simply a fact of life in this high-risk, high-stakes industry. Through decades of working with distressed and failing construction businesses, we can confirm that in 99% of the cases, the problems resulted from management decisions and management decisions alone. Construction professionals need to understand that the production side of an organization is completely separate from the business side of an organization. There isn't even any overlap.

The business side is *exactly* the same for a road builder, electrician, glazier, or building contractor. Managing the production and business sides of the organization requires two totally different skill sets to the point that they are like two different hats, and you can't wear both hats at the same time. Once you start wearing one hat at a time, your sense of control will improve by orders of magnitude.

2.6 Business Planning Strategies

Properly managing the important day-to-day marketing, production, and administrative areas of the business is not enough to ensure long-term success. That is because, in the construction business, short-term success in no way implies longer-term prosperity. Without forecasting and planning, a business can be driven in the wrong direction. In today's low-margin, highly competitive construction industry, it is imperative to address the longer-term goals of the corporation in order to ensure the continued well-being of the business. To carry the driving analogy a little further, contracting businesses have no reverse gear, so it's important to maintain a steady course in the optimal direction. If you drive too far in the wrong direction, you can't simply back up and restart because once you have committed your resources, your organization is propelled in a specific direction. The hiring and training of tradespeople and supervisors with certain skills and experience, along with the acquisition of construction equipment and the establishment of a reputation, are investments propelling the company in a certain direction. Changing that direction, even slightly, is extremely difficult and expensive. For companies feeling compelled to correct direction, it is often too late.

Business planning is the single best and most effective defensive tool, and for that reason alone, all businesses benefit from planning. Effective planning also provides a framework for making better decisions throughout the organization by identifying future opportunities to be exploited and threats to be avoided. Business planning provides guidance to leaders, managers, and workers and aids them in making decisions in line with the goals and strategies of top management. Business planning also helps prevent piecemeal decisions and provides a forum to test the value judgments of the organization's decision-makers. Perhaps the most significant benefit of a well-organized planning process is improvement in communications among all levels of managers about goals and strategies to achieve the intended outcomes.

Properly performed, the planning process creates a communications network within even the smallest of companies that gets people excited about what's right for the company and how to achieve it. Planning also addresses an area that's sadly lacking in most small businesses today, which is the measurement of success. Establishing even a fundamental level of corporate planning in the smallest to largest contracting businesses has had profound effects on the outlook, attitude, and performance of employees and business owners alike.

One of the greatest selling points for comprehensive business planning is that it allows construction professionals to simulate the future and document it on paper. During planning, if the simulation doesn't work out, it can be erased and started again. With a planning exercise unlike trial and error, decisions are reversible, and ideas can be tested without committing resources to them and without betting the entire company on the outcomes. Simulating various business scenarios allows, even encourages, managers to evaluate and experiment with many alternate courses of action. Something that can't happen in the actual operations of a business.

Participating in the business planning process can also be incredibly empowering for individuals in an organization. While it is the purpose of the business plan to communicate which direction the company is going in, it is just as important for management to understand what direction the company should *not* be going in and why. Running these planning simulations helps managers gain a deeper understanding of the reasons behind a particular approach. This is particularly important as the simulation points out what not to do. Not only will participants feel more committed to a plan they helped develop, but they will also be more effective at implementing the plan with the knowledge about the types of things the company should avoid doing.

The planning process brings more and better-tested information to the table to consider when making decisions. The ability to experiment with different courses of action without committing resources encourages planners to stretch their creative skills in a safe environment, resulting in ideas that may otherwise have been missed. Models of real-world situations provide the opportunity to test different scenarios and to evaluate potential consequences. This process makes selecting the *right* course of action more apparent and prevents costly mistakes.

No one can predict the future with 100% accuracy; however, the probability of certain events having a predictable cause-and-effect relationship can be evaluated fairly well. The more you know about your business, your marketplace, and your competition, the greater the likelihood that you can simulate the outcome of various moves. Planning is a remarkable tool that provides a competitive edge because the longer a company has been planning, the better it gets at projecting the outcomes of decisions and exploiting opportunities before competitors do. And of equal value, planning reveals threats to the business.

The days of assuming your business is helpless in the face of market forces are gone. Rather than reacting to market-created booms and busts, planners embrace an approach that allows them to determine their future by establishing their objectives in advance. If they are not met, planners will at least know the exact reasons why and will not repeat those mistakes again.

2.7 Overhead

The activities discussed so far in this chapter affect operating costs – corporate overhead. Managing overhead costs includes properly structuring an enterprise, establishing an efficient chain of command, and managing ongoing operations. The obvious question is, how do all of these activities affect overhead costs? Overhead has always been a controversial subject in the construction industry. Indispensable management, marketing, administration, and accounting expenses are costs of doing business and are essential support of the production side of the business. Establishing the precise balance between too much and too little of these costs is a critical and yet complicated pursuit. Too little and you risk underperformance or worse; too much and you unnecessarily waste profits.

Overhead plays a significant role in the performance of every construction enterprise. Some contractors see overhead as a pure cost center that is somehow negative and therefore needs to be minimized to the extent possible. This view results in underperformance because of a lack of management resources. Others have gone overboard with overhead as a control process, resulting in costs so high that underperformance results. Overhead is neither good nor bad. As a necessary and appropriate cost of doing business overhead is neutral. It's also complex, particularly with the continuous-growth culture that is prevalent in the industry. There's even a construction industry saying about growth: *"If you're not growing, you're going backward."*

2.7.1 Incidental-growth Pattern

Another often-heard fallacy in the industry is the belief that a company can do more, maybe even twice its current sales, with the same amount of overhead. The reality is that in just about every case, as soon as sales increase, overhead increases. This cause-and-effect relationship is logical. A problem that arises in managing overhead costs is that even modest growth often causes the need for more employees and equipment, but you can't get half an employee or half a copy machine. This forces companies to increase their overhead costs more than is required for the modest growth that has occurred. Consequently, companies feel compelled to go after more growth to support the increase in overhead. Consider the following example: A construction company captures two larger projects within months, and the increase in sales requires an increase in overhead. Before the two projects are completed, the increase in overhead requires an increase in sales (i.e. more growth). This cycle fosters growth as a continual necessity, and this pattern has become the norm among many companies.

This pattern can and does lead to what we call an *incidental-growth pattern*. Some people might prefer the term *coincidental-growth pattern*. The pattern continues because contractors almost never consider going back to the size they were before the unexpected two larger projects. Even in cases where growth was not planned, growth often initiates a pattern of continual growth.

A pitfall of continual growth occurs when small and midsize companies transition into larger companies because those transitions require a major shift in structure and major increases in overhead costs are necessary. As described in our prior discussion of construction company stages of growth. In a small operation, the contractor personally runs the work. In a midsize operation, senior managers run the work. Switching to senior managers running the work significantly increases overhead costs immediately because their costs are not usually able to be covered by current sales, and sales can't be increased over-night. This transition always causes a reduction in profitability and often results in losses that are close to impossible to avoid.

All midsize and large construction enterprises have gone through this transition even though they had no way of recognizing the cause because the mechanism was only recently recognized. The timing of the transition varies widely between general contractors, construction managers, and subcontractors because the timing is affected by the type of work they do. Therefore, total sales isn't a good indicator. The point at which a small company becomes a midsize company is when the contractor/CEO no longer personally causes the work to be produced. The point at which a midsize company transitions into a large company is when the contractor/CEO no longer personally directs the managers responsible for producing the work. As described previously, jumbo firms are just groups of large firms.

2.7.2 Overhead and Organizational Structure

As mentioned earlier, construction companies in the United States have been managed in hundreds of random ways typically selected according to personal choice. In the past, when organizational behavior, structure, and overhead management in the construction industry were studied at all they were considered separate components. Although managing overhead is part art and part science, overhead in construction is so intricately entwined with organizational behavior and structure that it can't be studied separately. Research on construction industry overhead must consider organizational behavior and structure in order to be of any value. This relationship has been ignored until recently to our industry's detriment.

The relationship was questioned during three years of research into construction industry overhead. When organizational behavior and structure were introduced into the study, it was recognized that how a construction organization is structured and operates were the driving factors for the configuration and composition of overhead. Once we realized that overhead was *responsive* to operations, or you might say that overhead is a byproduct of operations, the direction of the research dramatically changed.

When the incidental-growth pattern is in play, overhead typically is a driver of sales and indirectly of operations. This relationship caused more than a year of additional study because it was such a dramatic departure from prevailing beliefs, processes, and behaviors. The pattern indicated that organizational behavior and corporate structure dictated overhead requirements. Understanding this relationship helped to answer a lot of previous questions. The relationship between overhead costs that are permanent and sales that are variable seemed like an improbable business model. Overhead employees are permanent, full-time employees. Office space ownership or rental is long-term. And office furniture and equipment are not returnable.

This meant that sales must be adjusted to accommodate or cover any increase in overhead, which would cause incremental growth. Or: Overhead would have to be adjusted to accommodate any change in sales. This was because it was believed that adjustments (increases) to sales were within the control of the firm, and overhead was less so in that added people, equipment, etc. were in whole increments – no half person or have a piece of equipment. This seemed to explain why continuous growth was the goal of most construction organizations. Continuous profitable growth of course being almost impossible in a cyclical market.

2.7.3 Flexible Overhead

What this suggested and seemed impossible was that if overhead needs to be responsive to organizational behavior and structure, then overhead can no longer be treated as a permanent, fixed cost. The challenge was to determine if somehow overhead in the construction industry could be impermanent. As it turns out it can, but it took another year to develop the solution. In spite of the difficulty and what appeared to be an unfeasible task, *Flexible Overhead* was conceived and developed.

Flexible overhead is an important concept for construction professionals who want to be successful in good times and bad – whether there is a lot of work or there's not enough. The old paradigm of *continuous growth* doesn't work in a cyclical market and should be substituted with *continuous profit*. Consequently, the construction enterprise of the future must be organized in a way that its annual sales can go up and can also go down in order to cope with market realities. In which case it would never again be forced to chase inappropriate work just to maintain volume.

The reality is that the continual-growth model may work in a healthy growing market. However, the continual-growth model results in industry-wide drops in margins and increases in contractor failures in a cyclical market when the market declines. The reason is that in a growth-driven industry contractors refuse to give up the hard-earned sales increases they enjoyed during good times, even when the cycle changes. Construction has always been a cyclical industry, and the cycles will continue. Construction professionals have no choice but to replace the long-ago-established drive for size with a drive for profits and for prosperity. The objective is not to maintain profit as measured in size but to maintain (to the extent possible) the percentage of profit earned from the work produced. In a declining market, this approach means settling for reduced sales with reduced profits while competitors break even or lose money.

How to implement flexible overheads is addressed in great detail elsewhere. The basics are

- Enable 15–20% of overhead costs to be turned off within a week.
- Use temporary personnel, interim office space, and so forth for some portion of clerical, administrative, and accounting functions.
- Use short-term rentals (e.g. trailers) (Figure 2.3) for some portion of office space during growth stages.

The modest added costs, if any, that result from flexible overhead are not unlike an insurance premium for protection from a known risk. Flexible overhead also prepares a firm to do up to 25% or more additional work for short periods without permanently increasing overhead. While the flexible overhead approach is a departure from the current norm, the approach is part of the profile of the successful contractor of the future because an organization concentrating on profitability and not size has a different understanding of the market and the organization's place in it. Maximizing profits, not sales.

Of course, we're not saying there's anything wrong with revenue growth as long as it doesn't sacrifice productivity and efficiency and the organization understands that growth often lowers profit margins temporarily. The focus should be on maintaining the organization's vision and following the business plan. Traditionally, construction companies have striven to keep all their overhead during slow periods, and some owners have believed that because it took so long to find and train their staff, it would be impossible to replace them when the market rebounds. However, the real question often turns out to be whether cuts should have been made earlier.

Figure 2.5 Job trailers can provide temporary office space during growth stages.

When work slows down, the worst response is to go after projects that contradict the company's goals and don't fit the company's core competencies. A better option is to slow down sales and reduce overhead costs, which is easy to do with the flexible overhead approach. Rather than joining the feeding frenzy of reckless bidding to maintain sales, it is far more prudent to concentrate on what work you can capture and to shrink your geographic area to stay closer to home, allowing enhanced supervision to increase efficiency and profit. Leadership is about growing and enhancing people, not sales, and after the construction industry slows down, it eventually rebounds. How soon? Historically, in one to three years.

If one to three years sounds like a long time, it is even longer if you're breaking even or losing money while trying desperately to maintain prior sales volume. It's so much easier and more satisfying if you're able to do less work, allowing sales to reduce while profit percentage maintains or drops minimally. Willingly reducing sales may be a unique concept to some readers, but compare it to the alternative of putting work in place for one to three years for very limited profit or at a loss. Contrast the risks associated with each alternative. With the latter course of action, there's the promise of increased risk, including losing money, which, as discussed earlier, can trigger serious reactions from your bonding company and/or bank. In this scenario, you are in reality risking your entire company. The fact that you don't see the situation that way doesn't change the reality that it has happened to hundreds of other contractors. Our industry's history documented this clearly.

Flexible overhead facilitates the reduction of sales, allowing a contractor to cooperate with a market slowdown. Targeting a smaller profit. But nonetheless, a profit presents a limited risk. In most cases, the risk is just the regular risk of being in the construction business. This approach defies the popular industry belief that we have to grow or we're going backward. The contractor of the future will replace that fallacy with the belief that we have to profit or we're placing our business at risk.

(As a side note, when one of us explained at a seminar that the primary purpose of a business is to produce a profit, a participant asked, "Then what do you call an operation that loses money?" Our reply was, "*I'm not sure, but I wouldn't call it a business because it won't remain in business very long.*" Another participant said, "*What about a nonprofit business?*" Our reply was, "*The expression is an oxymoron. A nonprofit business is a charity.*")

2.8 Supporting Production in the Field

None of the information so far in this chapter is intended to distract from the primary objective of a construction business, which is and must be the production of the work at a profit. Accomplishing that objective at world-class performance levels maximizes a fair and reasonable profit. Of the three functional areas of business, production is the only functional area capable of producing cash flow – and cash flow is the fuel that a company operates on. Therefore, the other two functional areas, marketing and accounting, while equally important is supporting production, which is sometimes referred to as *project management*.

This relationship is significant, must be understood by all overhead staff, and when not understood leads to inefficiencies. Project managers are recorded as an overhead cost in some firms and as a field cost in other firms. In a functional sense, project managers produce the work and, along with field superintendents, interface with office, and other employees. Serious inefficiencies occur when office employees are not aware of field employees' contributions to the success of the company and believe that office functions take priority over field functions. We have experienced office personnel who consider field phone inquiries and interactions as interruptions of important office functions and even a nuisance.

Consider the following examples: An accounting staff member told a superintendent inquiring about a payroll error that "*I don't have time for this.*" A receptionist hung up on a yard driver who asked for a project address. An estimator refused a call from a superintendent who wanted clarification on the pricing of certain work. A project manager told a foreman to hold his questions until a scheduled visit the following week. The yard and shop staff sent a notice that requests for company-owned equipment, supplies, and materials from the yard had to be sent in writing at least a week in advance. And more memorable than most, a printed sign at the home office entrance to the executive floor stated, "*No Field Personnel beyond This Point!*" We could go on for pages. This attitude occurs more often in midsize and large organizations than in small firms.

The point here should be clear. All company overhead staff – marketing, estimating, administrative, and so forth – of a world-class construction enterprise should be trained, fully understand, and enthusiastically embrace their primary role as supporting, assisting, enabling, and advancing field activities. We're not saying that the field is more important than the office or that there aren't times when a marketing activity or some other event can be interrupted without offending a client. What we're saying is that the field and the office have different roles in producing the profit and cash flow of the enterprise, and without the field, there is no cash flow.

This fact doesn't detract from the importance of the office functions; rather, it clarifies the purpose and priorities of the entire organization. Company leaders are responsible for training all personnel, particularly new hires, so they understand the firm's objectives and each employee's role in accomplishing the objectives. All of the office-centered organizations we have experienced underperformed financially. That says it all. Field operations is the only place where the company earns the right to invoice a customer. The primary objective of every employee of a construction organization is to support the production of the work.

3

Construction Business Risks

It should be apparent that this book is about managing a construction enterprise for a profit and that we believe doing so includes protecting profits and the contractor's investment in the company. In this context, protection includes recognizing all of the risks that a construction organization might face. There are risks too numerous to mention, and many that have been more recently recognized are hidden risks. Despite the decades of research on construction risks, we may not have identified all of them; a fact that causes us to be concerned that each time we uncover a hidden risk, it probably won't be the last one. Risk is a theme throughout this book, and the topic is critical enough that we've devoted an entire chapter to risk. Managing risk is a critical construction business skill set.

This chapter discusses the most common and some of the not-so-common risks that can adversely affect construction organizations. We offer methods to avoid these risks and to mitigate their impact when encountered. Awareness is the key to mitigation. You would think that because of the nature of the industry and its failure rate, all construction professionals would be keenly alert where risks are concerned. Unfortunately, construction professionals are typically too focused on how to do the work and engrossed with the production of the built environment to think much about risks. Even though industry practitioners know that construction is risky, it is not something many believe they can do anything about. Further complicating the issue is that when asked about construction business risks construction professionals generally respond with comments about safety, material handling, and the like. While these risks are important, they are field risks not business risks. Insurance carriers and safety personnel are regularly concerned about these field risks, but management risks and generic business risks are very much *out-of-sight, out-of-mind* in too many organizations.

For example, when we ask middle managers about business risks, they tell us that top management deals with business risks. Most busy top managers tell us that they rely on middle managers to flag any risks that are encountered and to send the information up the chain of command. Although everyone in construction knows it's a high-risk business, hardly anyone is doing anything about it. On the bright side, few are losing sleep over the subject, but on the negative side, it's no wonder that just about all business risks that affect construction organizations are described as hidden risks. We hope that this chapter may change that situation.

The Business of Construction Contracting: Schleifer's Guide to Financial Success, First Edition. Thomas C. Schleifer and Aaron B. Cohen.
© 2025 John Wiley & Sons, Inc. Published 2025 by John Wiley & Sons, Inc.

3.1 Definitions

Let's start with some basic definitions to ensure we're all talking about the same thing.

- Short definition of *risk*: To expose to danger or possible loss.
- Measurement of risk: The likelihood that the risk will occur and the magnitude of the impact.
- Cost of risk: The expenses incurred because of the risk occurring. These expenses may include Occupational Safety and Health Administration fines, governing body fines, emergency response costs, as well as other expenses.
- Insured risks: Someone else may pay for the negative results of the risk occurring, but the risk is still our responsibility to prevent.
- Uninsured risks: We pay if we don't prevent the risk from occurring. In business, we own all the uninsured risks, including unreported mishaps.

To fully understand risk, we need to elaborate on the short definition of *risk*. In this definition, the probability of the risk occurring and the result of the risk occurring are intrinsically connected. It's important to remember that risk isn't necessarily negative. It can be neutral or positive. In addition, the occurrence of a specific risk can be negative for one person and positive for another. For example, you get up in the morning, and the weather channel announces there's a 60% chance of rain. If you're planning a picnic, the risk of rain is negative. If you're a farmer, the risk of rain is positive.

The only way to measure risk is to consider the odds of the event occurring *and* the results if the event occurs. To clarify, let's contrast two potential events with the same odds and see how dramatically different our choices might be based solely on the magnitude of the event if it were to occur. First, are you willing to accept a one-dollar bet at two-to-one odds – say, the flip of a coin? Most people wouldn't be too concerned about making such a bet – there are reasonable odds, a reasonable penalty (lose $1), and a reasonable reward (win $2). Would you accept a bet at the exact same odds if the penalty and reward were different? How about if the bet were for one million dollars?

To drive the point home that both the odds and magnitude of the event must be considered in the measurement of risk, would you accept a bet with six-to-one odds in your favor? Most people would respond yes because six-to-one odds are pretty good, and they are thinking of the above one-dollar bet to win $6. How about if the bet you are offered is Russian roulette? One bullet is put into the cylinder of a six-shot revolver. You then spin the cylinder, after which you point the revolver at your head and pull the trigger. How much would you like to bet? A rational person wouldn't take the bet at any amount. The result or the penalty, if one out of six events occurs, is too severe to be attractive at any odds. It should be clear that the measurement of any risk can't be made only on the odds of the event occurring alone or on the value of the results alone. The measurement can *only* be made using a combination of the odds and potential results.

3.2 Awareness of Risks

As mentioned before, construction professionals, particularly those involved in field activities are keenly aware of structural risks, scaffolding risks, product risks, lifting risks, and so forth. Generally, top managers are aware of these risks because of field experience. In contrast, the countless risks on the business side of a construction enterprise don't get the attention they deserve. In our years of talking with senior construction professionals, they have focused on field risks giving little if any attention to business risks. For managers located in home offices, it's almost as if business risks aren't an issue. Therefore, when something negative happens, it's considered a one-time event that management should not have been expected to anticipate. Herein lies a serious industry failure.

The only way to manage business risks is to recognize them. In other words, you can't manage risks to your business if you don't know they exist. One purpose of this book, and the primary purpose of this chapter is to remind practitioners that this industry has the second-highest business failure rate in the nation, and 99% of the failures are caused by business risks not field risks. Obviously, there are more than enough field risks to go around, and fortunately, they don't always put construction enterprises out of business. While they do cost, most field risks are insured. In contrast, business risks can be disastrous and hardly any are insured. It is important to point out that business risk insurance shouldn't be confused with performance bonds, which are not insurance and protect only the project owner not the contractor. The same goes for payment bonds that protect only subcontractors and suppliers.

3.3 Business-risk Categories

The following are some of the more common categories of construction business risks:

- Financial loss
- Project selection
- General Contractor and Construction Manager selection
- Owner selection
- Project designer selection
- Subcontractor/vendor selection
- Availability of subcontractors/vendors
- Performance of subcontractors/vendors
- Long-lead critical and sole source items
- Availability of field and office management
- Employee selection
- Employees' skill levels
- Equipment availability
- Size of project
- Geographic area of project
- Type of project
- Change in key personnel
- Managerial maturity
- Reputation
- Characteristics of public versus private projects
- Union versus merit shop
- Labor disputes

- Project management issues
- Project scheduling issues
- Pacing of project issues
- Scheduling ability
- Liquated and actual damages
- Quality/workmanship issues
- Defective materials
- Defects by subcontractors
- Failure to complete a project.

We've heard construction professionals say that the business is so complex and difficult that they are too busy to spend energy pondering potential risks. Management has been accepting those excuses for decades. Just because managers are busy and the business is complex, doesn't mean that we are not responsible for learning about the risks, that we are not responsible for preventing the risks, and that we are not responsible for the consequences of those risks.

3.4 Risk Management

It is probably safe to assume that the readers of this book aren't out of business and that some of you feel strongly that information about the potential for failure doesn't apply to you. Other readers may think this chapter should be of interest to their weaker competitors. This material is applicable to all construction professionals, particularly those in very successful firms, because this chapter is really about how to *remain* successful in a highly risky business. We're hopeful that the word *failure* doesn't cloud the message about how to manage and mitigate risk in a risky business. All of the failed construction companies were successful prior to their failure, so it is fair to say only successful construction companies fail.

Risk management is everyone's responsibility (Figure 3.1). It is important that every office and field employee understand and have a working knowledge of business risks. It may seem an overstatement to say that each employee must also accept and embrace that responsibility, but doing so is critical. Risks exist on their own, are not about to disappear, and are ever-present in the here and now. These facts come as a surprise to many construction professionals because they perceive risk as something dormant. Something that you'll see coming if it approaches. Recognizing the reality that there are multiple types and magnitudes of business risk in construction is the biggest step taken in risk management.

You are probably familiar with the term *risk prevention*, but in some ways, it is an oxymoron. We can't prevent risks from existing because risks exist on their own. Risks are there whether we like it or not. The only thing we can do is reduce the likelihood of risk occurring. We can engage in *risk control*, which we prefer to label as *risk management*, which doesn't reduce the risk but reduces the likelihood of the risk occurring or it reduces the impact if the risk does occur.

The first step in risk management is recognizing that risks exist and that it can be lethal. The second step is taking precautions when warranted. The third step is mitigating the results if the event occurs. For this reason, we will highlight known risks that will therefore no longer be hidden. However, it's important to note that we can't be certain that new risks won't join the list.

Figure 3.1 In a construction business, risk management is everyone's responsibility.

3.5 Financial Risks

Business failure ultimately boils down to one cause: unexpectedly running out of cash (working capital). Construction contracts assign an inordinate amount of financial risk to the contractor. This imbalance in risk assignment has gone largely unnoticed to the point that it's become hidden in the *bones* of the business. As a result, too many contractors are constantly at risk of having insufficient cash flow. In fact, a shortage of or strain on working capital seems to simply be part of being in construction, but it also affects the ability to take on new business and grow with confidence and stability. Growth and expansion in any business require a serious assessment of the financial capacity of the company.

It seems that when contractors consider more or larger projects, they seldom question the financial risk. The prevailing attitude appears to be that if they have been able to finance the work in the past, they will continue to get by. And they do. Until they don't. Running out of working capital doesn't happen suddenly. It sneaks up on you over time as you go about the normal course of business. In construction, financial risk can increase even while profitability is maintained.

The fast pace of construction and the lengthy timeline of major projects complicate the detection of financial risk, particularly in a growing business in which growth tends to cover up poor financial performance. Measuring financial risk includes looking at past financial performance, the use of earnings to increase capacity, the capital required to complete projects under consideration, and the financial drain of ongoing projects. The availability of outside capital in the construction industry is limited, relying primarily on bank lending, so contractors must retain earnings to increase their capacity to fund growth.

The two major risk factors built into contracting risks are undercapitalization and highly complex transactions. Just about every contractor we've worked with over the years has been distracted by cash flow problems at one time or another because construction companies rarely retain enough earnings to accumulate abundant working capital. Public companies routinely sell stock to raise growth capital, but the majority of construction firms are closely held and reluctant to sell even part of their business. Since ongoing business for contractors is primarily *new* business, earnings are rarely stabilized and are often unpredictable. The situation is further complicated because cash flow requirements differ widely from project to project.

Banks seldom place full value on construction contract receivables. They often discount the total to establish the contractor's working capital line of credit. In addition, most banks secure lines of credit on the tangible net worth of the company and its owners. This is problematic because the net worth of a construction firm has little or nothing to do with the amount of working capital required to see a particular project to completion. In effect, contractors have to build the work and finance it with limited access to capital.

3.6   Inappropriate Industry Risks

Construction enterprises are service providers. However, we've inadvertently allowed the industry to evolve in such a manner that our clients treat us as product providers. We sign contracts with government agencies, private owners, and developers that require us to construct, finance, and pay for the work before we get paid. Even worse, we're not paid at the time of the sale or in advance of delivery. We're paid (supposedly) in 30 days or sometimes months after the work is put in place. The owner has legal possession of our product from the minute it's put in place because the owner owns or controls the underlying property our work is placed on. Our industry has allowed the construction transaction to become so one-sided and complicated to the point that signing a construction contract requires us to invest our money in our customer's asset.

To fully understand this situation, let's look at the typical sale transaction that other industries routinely execute. A firm conceives, designs, and manufactures a product for an identifiable market. The firm calculates the manufacturing cost and adds their overhead costs and a reasonable profit to set the selling price. The firm sells the product and collects payment for it (directly from the customer or a financial intermediary) before or as the ownership of the product is transferred to the buyer. The manufacturer (seller) can then reinvest part of the payment in ongoing manufacturing activities to meet future demand and can retain some profit for other purposes.

Let's contrast this process with the typical construction sale transaction. Our client conceives and designs the product (road, building, utility, etc.) that the client needs or wants. The client then invites contractors to compete with each other and bid on constructing the project. A process that some have referred to as a *Dutch auction* because the process requires potential service providers (contractors) to estimate the lowest possible price for their work. Contractors then estimate future costs in order to arrive at the lowest price they can provide their services for. In many cases, the contractor then has to try to determine how much to discount the price to capture the contract.

As if this process isn't one-sided enough, the client then requires the successful bidder to sign a legally binding performance guarantee drawn up by the client's lawyers. The contract document is included in the bid package, rendering the terms non-negotiable thereafter. The bidders guarantee that the project will be completed within the projected time frame and for the estimated price. At this point, many con-tractors begin to wonder if they missed any costs fearing estimating mistakes. There is of course no down payment, thus requiring the contractor to finance all costs until conditional progress payments are received. In other words, the contractor isn't only providing construction services but is also providing banking services in the form of financing the progress of the work throughout the term of the project. (A loan that often runs up to tens of millions of dollars.)

The contractor's risks are magnified when there is (and there often is) a contractual requirement to further ensure the contractor's performance through the provision of payment and performance surety bonds. These bonds, in effect, guarantee that if the construction firm doesn't perform in accordance with the contract, the surety will finish the work. Compounding this risk, the contractor is, in most cases, required to indemnify the surety corporately, making the construction firm financially responsible for any loss incurred by the surety. And in many cases, personal indemnity is required making the contractor personally responsible for any surety losses. Expanding this risk, most personal indemnity agreements require the signature of the contractor and his or her spouse if there is one.

As if these guarantees aren't enough, most construction contracts require the owner/client to pay only 90% of the contractor's monthly billing, retaining 10% of the money already earned by the contractor to further guarantee the contractor's performance. This practice is punitive at best in an industry in which contractors don't typically make anywhere near 10% profit. Owners and designers have been known to claim that this practice is justified because contractors front-load their requests for payment. Claiming that contractors inflate their monthly payment requests by placing inflated values on portions of the work or claiming that more work was completed during the pay period than actually was. (*In the interest of full disclosure.*) Historically, there may have been some truth to these arguments, but there is much less today because designers who approve quantities for pay applications have better data than in the past and have improved methodologies. Either way, inflating values up front does not work for long because later portions of the work have to be deflated by the same amounts.

3.7 Risk Identification

Early detection of financial weakness is critical to managing risk because detecting the financial weakness early provides management time to take defensive action. Contractors need a warning system that identifies any changes in performance. The system should be easy to use and not too labor-intensive. Further, the system needs to measure change by looking at a variety of components because profit margins, assets, liabilities, and debt are interrelated. So, measuring the changes in these relationships by using just one performance efficiency or debt ratio may not accurately reflect overall financial risk. For example, it's considered positive when the efficiency ratio of sales to total assets increases because an increase indicates the company is doing more work with fewer assets. However, the assets also support bank and bonding credit, which limits future needs or growth, creating future financial risk. The same thing occurs when liabilities increase and assets do not because the total-liabilities-to-total-assets ratio alters, and credit limits may be impacted creating a concern or financial risk.

Credit grantors vary in what they consider to be the most important financial ratios. Some creditors have focused on the relationship or balance between the turnover ratio of a construction company and its debt structure and consequently may limit the extension of credit. Lenders have limited or cut off companies with rapidly increasing debt causing financial distress and even failure if an alternate source of financing can't be arranged quickly. Because construction professionals' decisions and activities can affect financial ratios, it's important to be aware that performance efficiency and debt ratios are used externally to determine the financial strength (or weakness) of an organization and can be used to gauge the direction in which the firm is heading. The ability to recognize and measure financial risk is critical to the management of financial risk (Figure 3.2).

Corporate decisions and events that commonly cause construction company financial distress and failure usually precede the actual failure by two or more years, and modest changes in financial statements and in standard financial ratios are too often missed as indicators of an increase in financial risk. The fast pace of a construction business and the fact that major projects can last for years make it difficult to detect financial weaknesses and risk early on. This difficulty is especially present in a growing business because growth tends to cover up poor performance. The difficulty is reinforced by the reality that construction professionals all seem to be optimistic, being certain that their businesses are doing well. It's difficult to recognize risks when you don't think they exist.

Contrary to current belief, vigorously assessing risks in this industry makes a lot more sense than taking an out-of-sight, out-of-mind approach. What construction professionals should be doing is actively looking for early warning signs that financial performance is deteriorating. These signs need to be identified well before normal financial reporting or financial ratios display a deterioration in performance. Recognizing the signs early on will allow time to not just prevent failures but also prevent loss or any deterioration of profit. Employees should be responsible for managing and mitigating risks, not just constructing the work for a profit. It's time to set aside the belief that risks are unavoidable and inherent in the construction business.

3.8 Profit Versus Value

Dennis Borsch, a well-respected construction accountant, argued that *many US contractors are watchers instead of managers of their financial affairs.* He claimed that *they* (contractors) *do a lot of responding to things that happen to them but don't have much control over what's happening or going to happen because they don't have the appropriate financial information in the right place at the right time. To manage financial affairs properly, contractors must regularly analyze their historic financial data and gain a rearview perspective by carefully reading their financial statements.*

Figure 3.2   The ability to recognize risk is critical to the management of risk.

Financial statements are usually analyzed to measure financial performance, but we contend that financial performance can't accurately be measured without concurrently evaluating and measuring financial risks that may exist. People have conflicting views about whether financial performance should be measured by profit or by an increase in the value of the firm. We contend that the ultimate measure of performance is not what is earned, but how the earnings are valued by investors. A firm's increase in value (after earnings) is the true measure of performance. However, any business risks that exist affect the security of the value, and therefore, it is necessary to measure the risks to the firm's value in order to measure the firm's overall financial performance.

For example, consider two similar construction companies that have roughly equal values. One of the firms has little debt and a positive cash flow, whereas the other has increasing debt and cash flow problems. The financial risk (weaknesses) of the second firm renders it less stable and, therefore, of less true value than the first company. This difference in value would become apparent if they were both for sale, but absent such a transaction, the diminished value of the second firm may simply go unnoticed and unmanaged, allowing a bad financial situation to persist and potentially worsen. Any risk a firm has, including pending disputes or multiple claims, reduces its value. If and when the risks are settled, the value can return.

3.9 Financial-risk Measurement

In the up-and-down environment of the US construction industry, it's important to measure a company's financial performance and financial risk using financial statement data. The large number of construction business failures is indicative of the tremendous need to integrate accounting theory, financial analysis, and accounting data in a comprehensive performance measurement tool designed specifically for construction enterprises.

For this purpose, we have developed formulas that can measure changes in the performance of construction companies (Figure 3.3). One formula uses a combination of standard ratios in multiple categories to measure performance. The formula enables both financial and managerial organizational performance to be considered in an overall company evaluation. In the formulas, *organizational performance* is defined as a company's ability to produce its work at a profit, which for our purposes is the gross profit margin. Financial performance is a company's ability to remain in business by using its sources of funds.

Both financial performance and operational performance affect the efficiency of management and the use of a firm's assets. Sources of funds are used to acquire assets and efficient use of those assets is a measure of operational performance. Therefore, an overall company evaluation must address both financial performance and operational performance.

3.10 Operational Versus Financial Performance

Operational performance (gross profit) relies primarily on the organization's skills, systems, and the quality of employees. These components are supported and funded but not directly affected by the capital structure of the company. Because operational performance must be funded, it can't exist without positive financial performance. Therefore, any financial distress immediately affects operational performance. Otherwise, operational performance is independent of financial performance. On the other hand, financial performance is totally dependent on operational performance for funding through profits earned. For example, net profit (i.e. total gross profit minus general and administrative expenses) depends on operational performance. General and administrative expenses (executive, accounting, marketing, human relations functions, and the cost of capital) support the infrastructure to get and do the work. For most construction firms, the majority of general and administrative expenses are fixed costs that continue whether or not the company has any work.

Poor financial performance affects operational performance funding when working capital decreases for any reason, such as owners taking too much money out of the company or overinvesting in assets. Working capital is often a function of available credit, and if it dries up, performance is immediately

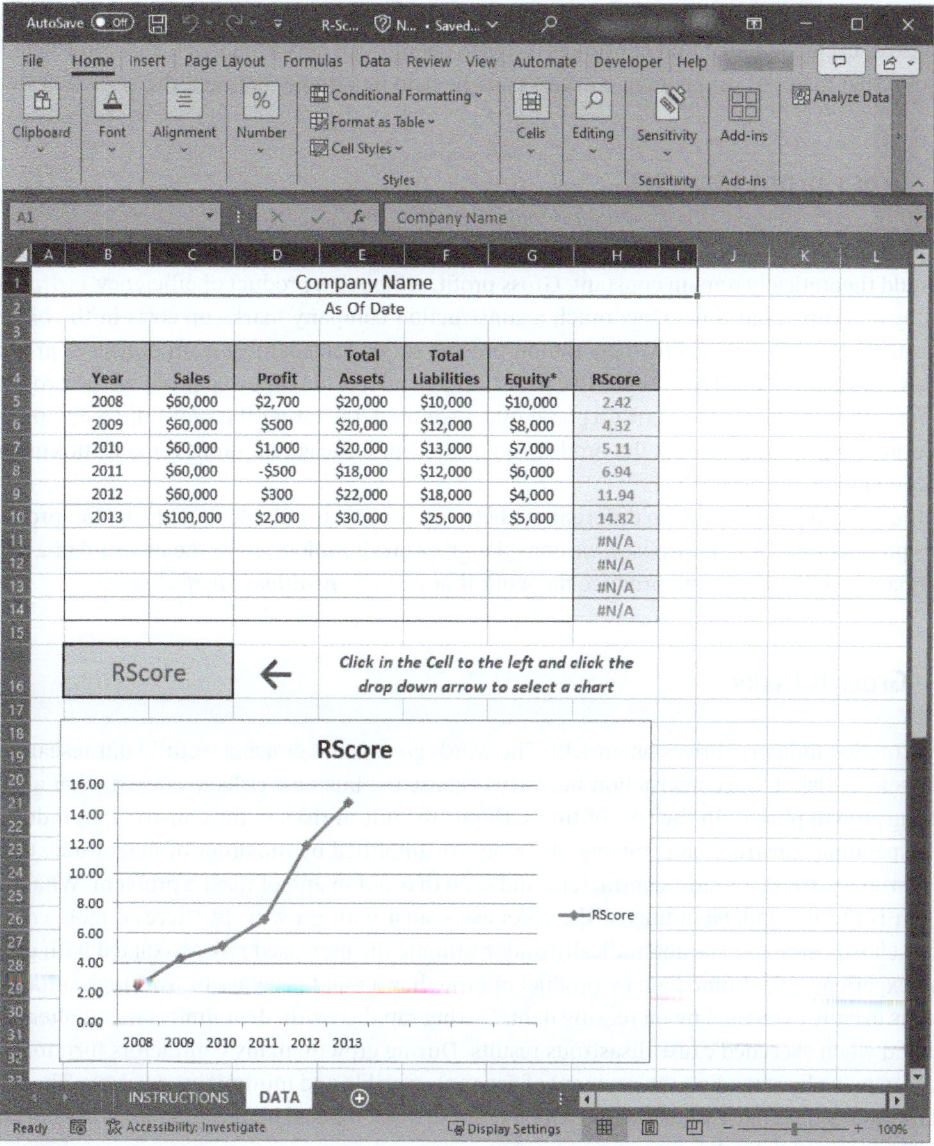

Figure 3.3 Example of a tool used to measure financial performance. R-Score Calculator, available for download from https://simplar.com/wp-content/uploads/2021/11/R-Score-Calculator-1.xlsx.

affected as evidenced by slower payment to subcontractors and suppliers. If accounts payable is extended as an alternative method of financing, then the subcontracted work and vendor deliveries usually slow down.

Reductions in working capital and the extension of accounts payable are usually the results of diminishing gross profit. Profit is the only risk-free source of funds in the closely held company. Other sources of funds such as financing are forms of risk. When poor operational performance impacts working capital, a negative cycle is created that continues to erode operational performance. If cash flow deteriorates to the point that material suppliers or workers can't be paid in a timely manner, the work will stop.

3.11 Gross-profit Changes

Although total sales for a construction firm will vary from year-to-year gross profits as a percentage of sales should theoretically remain constant. Gross profit, although a product of efficiency, is dramatically affected by markup at bid time. How much a construction company marks up costs in the bid process and adjusts their fee or profit in the negotiation process varies considerably from project to project. The markup is always influenced by competitive pressures and affects the expected profit on the project. Most people agree that the larger the markup going into a contract, the larger the profit upon completing the work. Because many factors affect the production of the work, the actual profit remains unknown until project completion and acceptance.

It's difficult, if not impossible, to differentiate between profit attributable to markup and profit attributable to production. For our purposes, we consider gross profit (or loss) to be the measure of a construction company's ability to get and produce the work that we call *operational performance.*

3.12 Growth Risks

The construction industry thrives on growth. The words *growth* and *growing* recur in our research on the management of risk in the construction industry because the business risks in construction are magnified during growth phases. In the best of times, there are built-in risks in the construction industry. In a rapidly expanding construction company, the risks are amplified by the order of magnitude. Growth is clearly a primary driver for most contractors, and growth is not in and of itself a problem. What makes it a problem is the misunderstanding of the risks associated with growth. In surveys, most contractors dispute their exposure to risks and radically underestimate the increased risk associated with growth.

In our experience, the immediate by-product of growth is a rapid decrease in working capital and this decrease is usually overcome by increasing debt. During rapid growth, debt limits are too often not considered and when exceeded cause disastrous results. During growth, many contractors turn to a form of internal financing by extending the payment of accounts payable and minimizing accruals. These actions are often combined with bank borrowing, which tends to mask the real increase in liabilities and obscures the increase in financial risk.

For many construction enterprises with limited credit, the primary source of financing is retained earnings. In this situation, sales and assets can grow no faster than the retained earnings grow or the increased debt that retained earnings can support. Too many construction professionals think that increases in sales will automatically increase their bank line of credit. This thinking is incorrect and, depending on the firm's financial strength, may even cause a bank to reduce its line of credit. Too many borrowers fail to understand that top-line growth (sales) is of limited interest to creditors who generally concentrate on bottom-line growth. Attempting to grow without assured financing leads to growing beyond a firm's financial capability.

It's extremely difficult for a construction enterprise to project how much it can effectively perform and finance, but all industry professionals should understand that every organization has a limit. It's also critical to understand that both financial and operational stress increase in direct proportion to growth. This reality is further complicated by the fact that growth often disrupts or slows recordkeeping, and this slowdown can obscure problems for a long time. Even highly profitable work puts a strain on cash flow, and few construction companies can gear up quickly enough or solidly enough to hold profit margins during growth periods. This is particularly true during periodic fluctuations in the labor market, the supply chain, and delivery delays. If these challenges occur even in good times, what will happen if a slowdown occurs? Growth almost always negatively affects margins, particularly in firms trying to capture more than their share of the market. Too often, contractors are the casualties of a robust construction market because they grow at a rate that unknowingly exceeds the growth rate their firm's financial capacity can support (Figure 3.4).

Figure 3.4 Contractors often fail by growing too fast in a robust construction market.

3.12.1 Overhead

Growth risks include overhead costs that are difficult to control even when a company isn't growing. In a growing organization, managing overhead is a very real and hazardous problem. During growth, construction organizations are forced to incur higher overhead costs because more resources are needed in the home office to support the increased volume, but most of these resources are difficult to add incrementally. For instance, it's not practical to add half a person or half a piece of equipment. Extra overhead costs cause underperformance or create losses until the company grows into the overhead by bringing in more sales. This problem is magnified when lagging profits make it necessary to increase volume to cover the increased overhead, thereby putting the company in double jeopardy.

A growing organization is almost always attempting to increase market share. Profit suffers because it's usually necessary to make at least temporary price concessions in order to take market share away from competitors. Although you may not make a conscious decision to lower prices to capture more volume, that's what occurs. When bid prices suffer, it's usually for all of the work, not just the additional work that the growth adds. When a decision is made to grow, all the work is aggressively acquired. Often, the company ends up needing even more volume than originally planned because margins are reduced. This results in a downward profit spiral because when an organization gets stretched, there's little time for anyone to see performance problems coming, let alone do anything about it.

Too many organizations pursue growth without measuring performance until it's too late. Rapid growth puts a strain on a company's key people and systems, and sustained growth doesn't allow for a reasonable training period. Of even greater concern, continuing growth doesn't give an organization a chance to test new people or systems before the next new people and systems are added. If performance deteriorates as a result of growth, the deterioration will be discovered after the additional work and people are taken on. Corrective measures are more difficult when people and systems are stretched out and when overworked managers are coping with the largest volume the enterprise has ever handled. Some companies don't recover from planned attempts to grow. No matter how hard you try, you can't measure the impact of growth on your organization's finances or performance until it is too late.

3.12.2 Rate of Growth

In a reasonable market, companies will presumably grow at a reasonable rate. There is a relationship between growth and risk. Years of research indicate that the growth of a construction organization of up to 15% annually is sustainable and does not appreciably affect business risk. The risks during short intervals of growth differ from continuous or rapid growth. At 15% annual growth, a company will double its size in five years and triple in seven years. At 25% growth, a company will double in three years and triple in five years. And at 50% growth, a company will double in 20 months and will be 500% larger in just 4 years. Continuous growth exceeding 15% a year is high risk and over 25% is close to reckless. Business failure and excessive rate of growth are intimately related.

Growth always requires more resources, people, systems, and money. Successful growth depends on an organization's ability to find qualified people, put appropriate systems in place before expansion, and finance the increase in costs. The rate of growth obviously affects an organization's ability to bring in the required resources. Growth always places an increased demand on existing employees and resources.

Few construction organizations are known for having underutilized resources or *bench strength*, so the stress from growth is immediate.

Determining the limits of expansion is complicated, and some highly respected management specialists don't believe there is a limit. Our research on the failure of companies during rapid or excessive growth confirms the opposite. Any growth in excess of 15% annually presents risk, and at accelerated rates of growth, risk increases exponentially. Many well-known and admired construction firms have failed during accelerated growth. While some point to other reasons for failure during growth, the reality is that growth is dangerous, and that dramatic growth is lethal. Any form of growth has risk, and moderate and intermittent growth are the safest.

A construction organization can't just grow because it wants to because fundamental financial constraints limit healthy, sustainable growth. Managing growth requires a careful balance of sales objectives with the firm's operating efficiency and financial resources. The challenge is to determine what sales-growth rate is compatible with the realities of the organization in question and its marketplace. All companies have limits in their ability, available resources, and capital. While difficult to measure, each organization is capable of safely doing just so much. During periods of rapid growth, construction companies can become so changed that they actually become new, untested organizations. This happens at a time when they have a lot more work to produce. The prior organization that was so successful is gone. It has been diluted by new people, processes, and projects.

The only reasonable test is for the organization to operate profitably and smoothly for a minimum of a year after excessive growth. If the test results are satisfactory, additional growth may be continued for another year. The same risks are present, and testing is required again. At the very least, tests are prudent. If testing is unsatisfactory, the organization has time during no growth to regroup, or it can roll back to its proven size. For companies involved in continuous growth, it's always too late to retreat and recover. The reality is that there is a limit to expansion. The problem is that the most common way to identify that limit is to exceed it.

If the expanded organization is to succeed, it must preserve its performance and quality as it grows larger. Healthy organizational growth doesn't dilute quality. It takes steps to maintain quality that needs to occur before sales growth not after it. The reason is that it takes more time to expand management and systems than it takes to capture more work. The more common but very high-risk approach is to expand management only after additional work is on hand. Growth just for growth's sake is risky in any industry, and growing in the construction industry is far more complicated than is commonly believed.

3.13 Management of Growth Risks

Incremental growth instead of sustained growth may seem unnecessary, even unnatural, but it's the best way (and very likely the only way) to manage the inherent risk in growth beyond 15% a year. After a short cycle of growth, typically a year or so, it is prudent, and we contend essential, to test the operational and financial results. If they're satisfactory, then another growth and testing cycle can proceed. An organization needs to evaluate periodically so that it can recover after a bad test. The alternative is to pursue continual, constant growth until that proves unsuccessful. The exposure is to wait too long, which is often too late for the company to recover. This growth and testing cycle is appropriate *risk management*.

During rapid and/or sustained growth a company can grow beyond its staffing systems and other resources so often that the company is never the same long enough to truly test the new status of the company. This situation can be described as functioning at constant risk with an ever-changing team. In many cases, failure is just a matter of time. Risk management requires that contractors grow cautiously, testing as they go, and to be prepared to withdraw from bad decisions.

Growth risk is magnified by the reality that in the construction business, growth eats cash. This is because construction enterprises put the work in place and wait for their money. During growth, a company is putting more work in place and invoicing for it each month than the month before. The company is paying out each month more than they are collecting from the prior months' work, so it will eventually run out of cash and credit. It is critical, though difficult, for contractors to accurately project how much work they can effectively perform and finance. A simple guideline is that if annual growth approaches 15%, it will invite increased risk. This demands serious consideration about the additional work. Can you identify the internal or external financing that will be used to support the growth?

Managing a closely held construction company is like driving a truck up a hill (Figure 3.5). The steeper the grade, the more strain on the truck, the engine, the suspension, and the drivetrain. These are the equivalents of employees, systems, and finances in a construction company. A truck that starts on a level roadway will have an easier time moving up the hill than if the truck starts already on the hill. Similarly, short separate growth periods are much easier than sustaining continuous growth. We've all seen trucks traveling up a steep, long hill slow down to a crawl. Laboring to gain forward progress, some actually stop.

Similarly, when a construction organization embarks on growth at a rate of more than 15%, the company's resources will be strained. During periods of continuous growth, the strain is sustained and magnified sometimes to the breaking point. Subjecting resources to severe and continuous stress encourages inefficiency (slowdowns), deterioration (crawl), and even failure (stop).

3.14 Market Recovery Risks

It will not surprise you to hear that construction business risks increase during a market slowdown. But it may surprise you to learn that the risks are even worse during market recoveries. At the beginning of a downturn, cash flow and the balance sheet look good as old receivables continue to come in, and less money goes out because less work is being performed month to month. The opposite occurs during growth. Spending on current work exceeds money coming in because the money coming in is for the lower amount of work invoiced for the prior month. A sustained downturn financially weakens most construction enterprises to the point that they may not be able to internally finance growth when the market recovers. During growth, the workload increases each month, and the outflow of cash increases to support the increased work. In this situation, it's possible for contractors to cash flow themselves out of business.

When contractors make it through the downturn, they usually believe they are out of harm's way. But the risks are actually magnified during a growth period following a downturn. This risk is especially dangerous for firms that resisted the downturn by taking on work at any price to keep the workforce busy while attempting to wait out the slowdown. Taking work with little or no profit reduces the financial capacity that will be needed when the market rebounds. Nevertheless, many desperate firms have accelerated their already weakening financial condition during a downturn market by fighting for every last

Figure 3.5 Managing a closely held construction company is like driving a truck up a hill.

project. Some to the point when recovery finally arrives, they are unable to finance their recovery or increase their line of credit.

The construction industry has cycled through a downturn on average every 10 years since World War II (WWII). Construction professionals can no longer claim that a market downturn caught them by surprise. The pattern during these cycles is always the same. During the downturn, the majority of top management refused to *cooperate* with the market and fought for every last project. There is only one way to fight and that is to lower prices. Profits of course go down. How much this weakens the company financially depends on the length of the downturn. When the recovery finally arrives, the financially weakened competitors continue the fight for the slowly expanding market, extending the distress. During the recovery periods, the failure rate has often exceeded the failure rate of the preceding downturn market cycle. This has been occurring since WWII.

3.15 Summary

The construction business requires a rare combination of talents, including the ability to gather resources, to build a project, to accurately price work in advance, and to manage workers, subcontractors, vendors, and designers through a long and arduous process. It also requires a tolerance for a high degree of risk. Historically, contractors of this breed had great success as long as their methods of bidding and their costs for items of work were closely guarded secrets, and as long as the production of complex projects remained a mystery. None of which is any longer the case. Decades ago, double-digit profit margins compensated for any and all inefficiencies.

All that has changed. Technologies and universal access to information have demystified the processes of estimating, organizing, and producing the work. Clients and designers have become quite knowledgeable about these processes and the underlying costs. Contractors now work harder with increased risks and reduced margins. Standardized efficiency and productivity have become the major differentiators among competitors. Continuous improvement has become not just a necessary cost of doing business but critical to survival. Recognizing and managing risk is not optional. Risk management is a critical and fundamental component of the job description of construction professionals.

Construction professionals should understand that it is extremely complex to envision how much their organizations can effectively perform and finance. Every organization has a limit. If a construction enterprise is growing at a rate greater than 15% per year, the company is at risk. The more rapid the rate of growth, the greater the risk. Consider carefully how additional work will affect your organization and cautiously approach growth. If you are concerned now, measure the extent of your present risk. You can do so by using the R-Score formula developed by Dr. Tom Schleifer. The formula is available for free at https://simplar.com/wp-content/uploads/2021/11/R-Score-Calculator-1.xlsx. Calculate your R-Score for the past three to five years and determine whether your risk is trending up or down. If the scores are rising over time, scrupulously examine your exposure.

In business, there is no such thing as no risk, and it is difficult to clearly define how much risk is *too much* risk. The definition of too much will vary from contractor to contractor because risk tolerance varies from person to person. It is fair to say that less risk is better than more risk and that risk management is as important as profit management. Risk management may even be more important in that too much risk can lead to business failure.

4

Common Elements of Construction Business Failure

The construction industry has changed dramatically over the years, becoming more sophisticated, employing new technologies at an astonishing rate, and providing services that are ever closer to being considered commodities. These changes are the primary reason that profit margins are low compared to historic norms, that risk has increased, and that there is less room for error. As mentioned previously, profit enhancement in the future will depend primarily on productivity improvements and efficiency.

Many construction professionals believe they lose money or fail because of labor problems, weather conditions, inflation, interest rates, increased costs of equipment, a tighter market, or simple bad luck. These factors contribute to financial distress once a poor management decision is made, but they are not the basic causes of construction business failure. Failure is not the result of factors or conditions that management has no control over. To survive, let alone thrive, in the industry, managers cannot let their guard down. Even natural growth resulting from ongoing operations involves change, and change has associated risks that can make or break an organization.

When a company expands in size, takes on larger or different kinds of projects, or enters new geographic areas, the company must make good management decisions to reduce the risks inherent in such expansion. An enterprise may be doing fairly well or even very well, but the stress that growth or change creates weakens competencies, sometimes to the point where those weaknesses become fatal. Change has to be managed to minimize risk.

How to manage a construction business is a topic that construction professionals do not talk frankly about. Construction professionals meet, converse, socialize, make jokes, and tell stories about each other but do not share how they run their businesses. These professionals never talk about mistakes that cost money, missed deadlines, or anything about business strategies because most of these professionals had to learn the essence of the industry on their own through personal experience and sweat equity. Once they gain that information, they consider it almost sacred and think that others should learn it the same way. Consequently, the industry has been reinventing itself every day.

In the United States, more people are involved in this industry than in any other industry. This includes the people involved in ancillary services and in manufacturing and transporting construction materials. There are more than a million individual construction businesses, and the number of these companies that go out of business every year is phenomenal. Construction may be the most competitive and high-risk industry in the nation. If construction professionals survive a mistake and learn from it, they

The Business of Construction Contracting: Schleifer's Guide to Financial Success, First Edition. Thomas C. Schleifer and Aaron B. Cohen.
© 2025 John Wiley & Sons, Inc. Published 2025 by John Wiley & Sons, Inc.

know something their competitors might not know, which is considered a competitive edge that they are not inclined to give away.

When information is not shared, the collective body of knowledge available to the industry is limited, causing significant lags in innovation and improvement in the industry. The industry needs to embrace the sharing of information about effective ways to operate and organize construction enterprises and to improve management and production. The industry also needs effective methods of announcing and controlling inherent risks so that construction professionals are not relegated to learning from their own mistakes. Mistakes that are often fatal to the enterprise.

In contrast to other major US industries, construction was late in the development of extensive training programs that educate personnel, from entry-level positions to top-management positions, and that are ongoing and under continual review and revision. Other industries, such as steel, oil, and banking, have such programs and also have systems of checks and balances regarding decisions and strategies, and have governing boards to ensure accountability and monitoring of management decisions and techniques. In contrast, most construction professionals learn how to run a construction enterprise by working with and watching someone who learned in the same manner.

There is a lot of truth in the old industry saying that *all you need to start a construction business is a pickup truck, a box of tools, a cast-iron stomach, a bad temper, and a forgiving wife.* Many construction professionals have started with less. Times have changed, and in today's market, there are only a few ways to run a construction enterprise successfully. There are still plenty of companies that have developed their own procedures and styles, but they cannot compete with world-class organizations. Some still believe that sheer energy, drive, ambition, know-how, and guts will be enough in this high-risk industry, which may actually work, but only for a while. In today's marketplace, these characteristics must be accompanied by organization, direction, planning, and objectives. Without an appropriate structure, proper organization, and risk recognition, success will remain elusive.

By way of review, a construction enterprise has only three primary functions – getting the work (marketing), doing the work (production), and accounting for the work (administration) (Figure 4.1). These functions are separate but equal in importance. To be dealt with effectively, they should be analyzed separately, with time and energy budgeted to manage each function appropriately. It is imperative that someone has direct responsibility for each of these three functions. This can mean three different people or two or one person that accepts personal responsibility for one or more functional areas. It is not unusual for a small enterprise to have one person handle getting and accounting for the work and to have another person handle operation. Neither is it unusual in a small company for one person to handle all three functions; nevertheless, the functions remain separate.

The person responsible for a function must be qualified and should be clearly recognized throughout the organization. Some owners may consider one function more important or significant than the other two functions and may put less-qualified people in charge of those two functions, but neglecting any of these functions is courting failure because they are equally essential to success. The individuals who are responsible for one or more of the three primary functions are key to the organization, whether or not they own a piece of the company. It is critical that the individuals believe they are personally accountable for the success of their functional areas. They are principals in the organization, regardless of whether or not they carry the title.

Figure 4.1 The three primary functions of a construction business are getting the work, doing the work, and accounting for the work.

4.1 Common Elements of Business Failure

Knowing the reasons why construction businesses lose money is the surest way to prevent unnecessary loss. Investigating hundreds of construction company failures has generated a significant body of knowledge on the subject. The events and decisions that precede failure have been categorized and quantified in order to identify the most common causes of failure.

One of the most interesting facts this research has revealed is that the events and decisions that cause or contribute to business failure take place during the company's profitable years. To look for the causes during the difficult years, when a company is losing money or merely breaking even, is to study the result and not the causes. In fact, it is easy to be misled when examining only difficult years, because operations during losing years can generate unusual events and decisions even if the contractor is unaware of impending loss.

As an example, a slow-paying client can create cash flow issues making it difficult for a company to pay subs and suppliers on time. As a result, the company may have larger-than-normal balances on accounts receivable with its vendors, with whom the company normally has a good working relationship. A natural response to the stretching of existing credit relationships is to begin establishing new credit relationships with different subs and suppliers. This expansion of a company's supplier network can easily be justified as "growing pains" or even seen as a sign of positive growth because the existing network of subs and suppliers proved insufficient to meet the company's needs. In actuality, the use of new subs and suppliers simply hides the true cash flow issues created by a slow-paying client. It also increases the company's debt ceiling and its overall debt. It also introduces a whole new risk to the production side of the business by employing new trade partners and utilizing potentially new and different materials on the company's projects.

Many companies struggle through several losing years before failing, so the profitable years, and therefore the years that should be studied, will be one to four or more years prior to the failure.

The research on the failing companies identified five common elements of business failure:

- Increase in project size
- Changes in geographic area
- Changes in type of work
- Changes in key personnel
- Lack of managerial maturity.

These are listed in order of significance. Project size is by far the most lethal element followed by geographic area, type of work, key personnel, and managerial maturity.

This chapter will explore each of these elements briefly, occasionally using general examples of how these elements affect an organization and its ability to make a profit. Subsequent chapters will discuss each element in detail, focusing on identifying and minimizing risks inherent in expanding businesses. It is important to understand that all of the decisions contractors made concerning these elements were consciously made and that the events were clearly recognizable at the time they occurred. And that they likely appeared to be routine business occurrences. The contractors who made decisions concerning growth, such as expanding into new locations or new types of construction, did not see these decisions as risky or dangerous. Indeed, with proper planning and controls, many of the decisions would not have been nearly as risky or dangerous, but failing to recognize that there is a risk or taking the necessary steps to manage the risk can prove to be disastrous.

We are not suggesting that a contractor should fear all growth and other changes. What we are saying is that at least one of these elements preceded the failure of a large number of contractors. Therefore, there is inherent danger in these elements. Also, having a complete understanding of the risks involved in these elements is necessary to maintain success. When two or more of these elements are present at the same time, they are without a doubt more lethal.

4.2   Increase in Project Size

By far, the most common element present in construction companies that failed is a dramatic increase in the size of the projects undertaken. The change to larger projects occurred during profitable years, and sometimes problems developed even before a larger project was completed. Undertaking larger projects is a natural part of growth for many construction companies. The magnitude of the growth is what we are addressing here. In some cases, the new projects were two or more times larger than the firm's previous largest project.

A company's profitability and potential for loss are directly related to the size of a new project compared to the size of the organization's typical projects and its prior largest project. The research confirms that if a construction enterprise operates at a profit when doing a certain size of projects and a certain top size of project, there is absolutely no reason to believe that the company will automatically profit if it takes on dramatically larger work.

The question is not whether an organization can take on a project that is larger than any the organization has previously constructed. Almost any construction firm can build a project two or three times larger than its normal projects. Rather, the question is whether a firm can construct the larger project at a *profit*. For example, if a company can construct one-million-dollar road projects or buildings, it can in all likelihood construct a two-million- or three-million-dollar road project or building. However, as the project size increases, so does the strain on the company's resources, technical abilities, and collective experience. This is true whether or not management recognizes this fact. So, again, the critical question is whether the company can make a profit on the project.

Whether a company can make a profit on the first job that is twice the size of its previously largest job is at best unknown and at worst unlikely. Making a profit on a job four times greater than the largest previous project would be close to impossible without additional resources and a huge amount of careful planning, neither of which is likely without outside help. Getting the resources required might be possible, but how will an organization lacking experience with a project of such magnitude even determine what additional resources are needed? Without previous experience, how can the company carefully plan the work so it is effective? Contractors that normally construct jobs of one million or one hundred million dollar, work in an altogether different environment than contractors that normally take on a three-million or three-hundred million-dollar projects. This principle holds true no matter a company's typical project size.

Consider the following example: a small firm typically has two or three major jobs at a time, each around $600,000 or $800,000, while also working on a number of smaller jobs. So far, the largest project has been $1 million. The company has an annual volume of $4 million and has a comfortable profit margin. Then available work decreases, and the company's backlog drops off considerably. A $3 million project is out for bid, and in desperation, the company goes after it and captures it. The company thinks its problems are over for a while.

In reality, the problems are just beginning because no one has considered the impact that a much larger project might have on the organization. Most previous projects took a year or less to complete, and one of the firm's large projects typically started about the time another large project was about to finish.

There was usually another good-size project about half completed. The company is not particularly conscious of its cash flow pattern because cash flow has not previously been a problem. The reason is that when a large project neared completion, the company had spent a lot of money without receiving the final payment. However, at roughly the same time, a project in its middle stage has generated substantial monthly payments and a large project is starting up and about to produce significant cash flow through front-loading. This pattern is common in the industry. It develops because considerable time and resources are required to capture large projects. Therefore, most organizations are limited to pursuing their larger projects one at a time.

Contrast the example firm's normal "balanced" cash flow with when they take on the $3 million job. At first, the front-load is terrific, but the retainage adds up quickly, and within six or eight months, the company's total accounts receivable will be higher than ever before. At the midpoint and toward the end of the job, the amount of accounts receivable will be staggering, strangling the business. And this larger project will likely take longer to finalize than any other project the firm has ever undertaken. Even though the project is similar to the organization's other work, the organization may be surprised at the level of inspection and supervision by the architect or engineer because larger projects are usually managed by more-senior inspectors. Municipal, state, and lender inspections often require more recordkeeping and reporting than the company is accustomed to, and that field staff can effectively handle. For example, OSHA and union work rules are usually more strictly monitored on larger jobs, and security and safety requirements, along with expectations, expand with project size and visibility.

Therefore, larger projects are not within the company's experience and expectations, even though the larger projects may be similar to other jobs the company has completed. The organization may be able to get the job done, but making a profit on it is another story and far from automatic. The situation is similar to a company that paves driveways and parking lots taking on an interstate highway project. Building a parking lot and building a highway are similar but certainly not the same (Figure 4.2).

(a)   (b)

Figure 4.2 Building a parking lot (a) and building a highway (b) are similar but not the same.

4.3 Expanding into New Geographic Areas

Another common element in construction companies that fail is the decision to expand to new locations (Figure 4.3). A contractor's primary area may be one county, half a state, 5 states, or 50 states. The primary area is the location the organization has normally operated in, is comfortable with, and has been profitable in. Changing the geographic area a company works in is almost as common an element preceding failure as is increasing the project size. There are many good business reasons for a company to expand into a new geographic area, such as a desire to grow the business, a lack of work in the primary area, and the pursuit of special opportunities. Whatever the reason, the risks must be recognized and

Figure 4.3 Expanding into an area beyond where a contractor typically works in can introduce risk to the company.

planned for. Again, the question is not whether the organization can build a project in a different location. Rather, the question is whether the organization can produce a profit working in a new location.

Most construction professionals assume that their type of work is performed the same way or in a very similar way everywhere in the country. This assumption is wrong. The differences in customs, methods, procedures, regulations, and labor conditions significantly vary across the nation. Organizations become accustomed to working in an area, and most are surprised at the differences they encounter when expanding into unfamiliar territory. Differences can be expensive, particularly when unexpected and therefore not planned for.

Examples are numerous: merit shop contractors bid outside their area without knowing that the work will have to be performed by union workers. In most areas of the country, underground pipework requires complete dewatering of trenches. However, in certain areas, underground pipework is accomplished in the water. (If you have never done it, don't try to estimate the cost.) In some areas, it is almost impossible to keep a full crew on the first day of deer-hunting season (a disaster if you have scheduled a continuous concrete placement). Regulatory and inspection requirements differ from one area to another (state to state, city to city, inner city to suburbs). We have even encountered areas where local suppliers reserve their best prices for local contractors or will only serve local contractors.

Without going into geological and weather conditions, there are enough differences between locations to cause prudent contractors to want to make certain they know what they are getting into when they take on work outside of their customary areas. Local assistance may be needed just to facilitate the permitting processes, to understand local labor issues, and to evaluate subcontractors and material suppliers. As if the risks related to a new location are not enough, contractors often compound the risks by taking on a much larger project than any they have completed in the past. The thinking is that it wouldn't pay to take on a smaller project so far away from their normal area. Therefore, it has to be a big project.

4.4 Expanding into New Types of Construction

Expanding into new types of construction is the third element that is common in companies that fail. For a variety of reasons, contractors expand the types of work they do, going beyond the work they have a proven track record with. Many were trying to add another type of construction to their existing specialty. A highway firm took on a sewage-treatment-plant project, and an earth-moving contractor bid on (and won) tunnel work. A building firm that had built one- and two-story projects went after a high-rise project, and a firm that constructs office buildings took on a hospital project. Of course, all are construction projects, but each requires different skill sets, expertise, and resources.

Contractors usually recognize the need to research and plan before taking on a new type of construction work. However, contractors often drastically underestimate the entrance cost – the amount that will need to be paid during the inevitable learning period as an organization adjusts to performing a new type of work. You cannot learn how to build work you have little experience with from books. And simply hiring a person who knows the new type of work is not enough. What usually happens is a company will struggle to complete one or more jobs before it can execute the new type of construction profitably. The unplanned-for costs can be extreme, and the distraction from other projects adds to the deficit. Unfortunately, some companies do not survive the attempt.

Most contractors are more specialized than they realize (Figure 4.4). Some construct several types of projects but capture mostly one kind. They may call it a coincidence, but it's more likely that they are better at bidding on and constructing that type of project. Organizations that are successful long-term usually start and remain with the types of construction in which they have expertise, and the organization's growth and achievement are based on the continued refinement of that expertise. Over time, they perfect their estimating for a type of work and become more competitive in getting projects. These organizations also get better at organizing the work and putting it in place, increasing their profitability as they grow. Ironically, many contractors that are well known for performing a certain type of work seldom identify themselves as being outstanding at bidding and building that type of work or recognizing it as their core competence.

Figure 4.4 Contractors are often more specialized than they realize.

Instead, many successful contractors tend to identify themselves with the largest, most complex projects they have ever completed (note we didn't mention if they were profitable). There may not be much excitement in the work companies are really good at and therefore those projects don't often garner a lot of attention. They don't have the same appeal as the bigger, more difficult, complex projects. The lesson is clear: being able to effectively plan and profitably execute the construction of a bridge does not mean a person can effectively plan and profitably execute the construction of a building.

It is possible for a company to begin performing a new type of work without even realizing they are making such a change. For instance, a more subtle change in the type of work involves changing from the public sector to the private sector or from the private sector to the public sector. This type of change has cost numerous firms a great deal of money, even when they haven't increased the size of projects or expanded to new geographic areas. True, a company can successfully expand to include private or public sector projects if the company has a healthy respect for the differences and risks involved and engages in some planning. Indeed, many companies do both public and private work and have been doing so profitably for years. We are not suggesting that companies should not cross over; we are just warning that many contractors do not recognize any differences in advance and therefore proceed to price and produce the work at a loss.

The following are typical differences between public and private work:

- Prebid qualifications to compete
- The criteria used to select a contractor
- The amount of collaboration between the contractor, owner, and others
- The quality of work expected and delivered
- The number and cost of changes assumed to be allowed for in the bid or to be contracted for during construction (change orders).

First, the qualifications to get on bid lists differ in the two sectors. In the public sector, bidders usually need to prequalify with the public body, the state, or an agency. In most places, any contractor can qualify with a little effort. Once the contractors have prequalified, they will have a good source of work to bid on because the bid lists are open to all contractors. They do not need to know someone to bid on a public job. The publicly available project lists are one of the reasons public jobs usually have more bidders than private jobs do. Initially, the project size a contractor can compete for may be restricted by the agency or by bonding requirements, but over time many start-up contractors are able to grow in the public sector.

On the other hand, most private sector projects have selective bid lists that are more difficult to get on because owners or designers pick the contractors to be included, often in an informal and less-transparent manner. Few start-up contractors can get onto the better private sector lists, and the number of proposers is usually smaller than for public projects of a similar size. The number of bidders on a project affects the results and the number of projects a company has to bid on in order to get one. Consequently, the cost of doing business may be higher, which negatively affects profit margins.

Most public bodies are required by regulation to award the work to the lowest bidder, whereas contractor selection in the private sector is often based on quality and price. The public awarding entity usually has limited control over the bid list or who gets the work. The various participants are often strangers, and the award and administration of the project are supposed to be kept at arm's length. A public project is usually administered "*by the book*." The contractor intends to perform according to the specifications and to do no more. The opposite is true of private work. The awarding party picks the bidders, may or may not open proposals publicly, and often ends up working with a known or preselected contractor. The owner, architect, and contractor are much more likely to cooperate and collaborate on a private project than on a public one.

Public projects are sometimes bid at a lower "*go-in*" price than are similar private projects, and some people may say the reason is that the number of change orders and extras is often greater on public jobs. The lower price going in on public work allows little leeway to perform even minor changes at no charge, whereas in private work with a team approach, minor changes are often handled informally without change orders.

On hard-bid public projects, change orders often enhance the modest profit the job was bid at. Private work is not priced as tightly because all parties usually expect a fair markup on the work and do not expect numerous change orders. The private project contractor also needs to preserve its relationship with the architect, engineer, and owner in order to obtain future work. Therefore, most contractors include a reasonable fee and profit in the bid price, anticipating the need to absorb the cost of minor changes. These contractors can then go about building the project, incorporating any incidental changes, while earning a fair price, and participating as a team member.

The differences in public and private projects can cause a contractor that has previously worked exclusively on public projects to bid too low on a private job. If the contractor gets the project and then expects charge orders for small changes, the designer and owner will think they are being mistreated because they are not accustomed to this approach. Consequently, the contractor will have unwittingly created an adversarial situation. A lack of knowledge about how the public and private sectors differ can result in constant and avoidable disputes because of differences in expectations. This situation is probably why select lists in the private sector are here to stay. This example illustrates some of the fundamental differences between the processes in which public and private sector contracting is performed. There can be significant differences in how the contract is administered from what the contractor is familiar with, despite the fact that the work itself may be essentially the same.

4.5 Changes in Key Personnel

The fourth common reason that companies fail is changes in the key personnel responsible for the three primary functional areas (getting the work, doing the work, and accounting for the work). In every successful construction enterprise, one or more top-level managers are responsible for the three functional areas. If a company is making a profit, it is primarily, if not solely, because of the efforts of these individuals. If one of them leaves, there is no track record of profitability for that functional area, and therefore the entire organization has no track record. The company is essentially similar to a start-up again. This simple reality is common in business, particularly for small- and medium-sized companies in the construction industry.

Some people will point to a business with six or eight good project managers and say, "*They're why this company makes money*." The same can be said about two or three key estimators who are primarily responsible for getting the work. Successful companies do not delegate responsibility for the primary functional areas to middle managers. The project managers, the estimators, and all other employees were hired, trained, and managed by the person or people responsible for the three functional areas or by processes these people put in place. Some people may think this statement is an exaggeration or an oversimplification, and others may think the statement is a slight to middle managers. The reality is that successful small and midsize construction businesses have a top-down management structure. (Large and jumbo firms may prefer a matrix structure.) The construction process moves too fast, is too complicated, has too many subcontractors and suppliers, and progresses to wait for or debate instruction. Similar to operations in the military, predetermined processes need to be rigidly followed and decisions made at the top need to be carried out with limited debate. Feedback, yes. Debate, no. (Readers who are devoted to the matrix structure may want to skip the rest of this section.)

The top-level management team at small and midsize construction companies, whether general contractors or subcontractors, is very small compared to the number of field people. The exceptions are construction management firms and some general contractors that subcontract all or most of their field-work. In both cases, the corporate organization is different than the field organization, because the two components have very different functions. The prime objective of the field side of a construction business is to produce the work at a profit. There are obviously other considerations, but they are nowhere near as important as the prime objective.

The prime objective of the management side of a construction business is to support the field side of the business. This statement may seem overly simplistic, but we state it to clarify that all of the many business and management functions exist to support the primary objective of making a profit. Profit is the lifeblood of a business because concerns such as maintaining a good reputation and complying with government regulation do not matter if there is no revenue. We are driving this point home because decades of research on the causes of construction business failures shows that many construction professionals have gotten caught up in the complexities of managing, planning, and executing and have lost sight of the primary objective.

When working with distressed companies, we observed marketing people who believed that getting the work was all the mattered, accounting people who believed that borrowing capacity was all that mattered, and administrative people who believed that accurate recordkeeping was all that mattered. While all of these activities are necessary, even critical, these people forgot (*or were never told*) that the primary objective of every employee is to help the company earn a profit. And the only functional area that generates revenue and profit is production. In an intense, complex business, top managers are responsible for teaching all employees about this objective and motivating all employees to focus on this objective.

The quality of field management depends primarily on the quality of the key top manager responsible for construction operations. If that manager leaves, the company will be permanently changed and will be at risk until the replacement proves that he or she can cause the work to be produced at a profit.

A key top manager should also be personally responsible for the firm's marketing, including pricing strategy, estimates, and final bid amounts. The takeoff and estimating staff may be a great asset to the company, but the top manager who brought them together is (or should be) ultimately responsible for the success or failure of capturing the work. If this key person leaves, the organization no longer has the team that has proven it can get the work.

There is frequently a "key person" problem in the administration function because it is regularly overlooked and underrated by most contractors. If a company has just two top managers (partners, for example) who are responsible for the three primary functional areas, one of them will be "*stuck with*" the administration function. Usually, this function is assigned to the person responsible for getting the work because administration mainly consists of office duties, just as marketing does. In smaller construction organizations, administration is often not even recognized as a primary area contributing to a company's success. Consequently, numerous small and even midsize companies assign the administration function to middle managers rather than to a top manager.

This problem becomes more acute as a small firm grows. When the business was small, the contractor could manage all the functional areas of the business, including details such as signing checks. Some contractors seem to come to believe they had learned accounting because they paid the bills, knew the bank balance, and dealt with borrowing. Administrative needs were few because the small contractor did not necessarily keep minutes during important meetings, confirm decisions in writing, or even reply to all of the correspondence received. The small contractor interacts with everyone involved in the business, so the impact of limited administrative duties is reduced. Commonly, as the company grows and the number of staff increases, administrative duties continue to be relegated to middle managers.

Assigning the role to middle managers works for a while, but the absence of a qualified person who is responsible for this primary function magnifies risk as the firm grows. The company will transition from being inappropriately managed to being inadequately managed. The company is certainly not organized for success. This situation can be compared to an army marching into battle with no one in charge of its supply lines. So, whether the key position is never filled or the person responsible for the function leaves the company, the firm is at risk.

4.6 Lack of Managerial Maturity

The lack of managerial maturity is another common element in construction companies that have failed. The term managerial maturity means that a contractor's managerial abilities mature as his or her business does. The contractor must transition from the principal of the company doing everything personally to building an organization that can do everything as well as or even better than the company's principal could do individually.

The lack of managerial maturity is perhaps the most widespread of the five elements of failure because this element is so often found in conjunction with one or more of the other elements. Some might say that this element may be a contributing factor in all the other elements. Many construction organizations were founded by one person. The entrepreneurs who survive the high risks of a start-up usually intend to grow their business. Having overcome the start-up risks, few realize they will be facing growth risks, because the qualities and abilities required for a contractor to succeed as a small construction business are not necessarily the same as those required for the success of a larger business. The self-confidence and independence that motivate entrepreneurs to want to run their own businesses often make self-evaluation difficult, which make it hard to realistically evaluate their shortcomings if any. It is hard for self-confident people to recognize the risks of growth and also to truly delegate responsibility to others in the business.

Many entrepreneurs think, *"If I succeeded at X volume, I'll do twice or three times as well at two or three times X volume."* However, at some point in the growth of every enterprise, the organization must change and become more sophisticated. At these junctures, more authority must be delegated, more complex systems and procedures will be required, and more sophisticated people may be needed to handle the systems and procedures. Most entrepreneurs instinctively follow a command-and-control strategy, which may be the best approach for a start-up but may not be the best approach for a growing business.

When true delegation is required, some command and control must be given up, and some founders have great difficulty doing so. It would be nice if these changes evolved slowly over the growth period, because they would be less drastic and easier for the contractor to digest. But the real world works differently. We cannot hire half of a person or put in half of a new system. It is all or nothing, which sets a pace many would like to slow down.

Knowing when and how to make organizational changes becomes a significant aspect of running a business and tests the entrepreneurial skills of the contractor in a growing firm. The organizational changes necessitated by growth, particularly major reorganizations, need to be made during successful times to ensure continued success. Changes are needed before problems highlight the need for change. The key to success in management is not to eliminate all problems but to focus on the problems of the present stage of the organization's lifecycle.

Those who resist change until they have proof of the need for change, such as having a losing year, have waited too long. Some of the organizational changes that are required so a company can increase from an annual volume of, for example, $5 million to $20 million or $50 million to $100 million, are difficult to recognize and may be even more difficult for some contractors to accept even when the required changes are recognized. These often include delegating responsibility and authority, hiring top managers to supervise long-time employees (e.g. associates, friends, and family members), and sharing financial information with more people. Entrepreneurs must embrace transparency, including sharing information with employees about the company's financial information.

Contractors who are unable or unwilling to change their organizations to effectively deal with growth must curtail their growth or will face the risk of the business outgrowing its organization. Attempting to do $100 million worth of business with an organization optimized to operate at a volume of $50 million is suicidal.

4.7 Summary

It is critically important to fully understand the five common elements of construction business failures in order to avoid them or at a minimum to plan for them. It is also necessary to understand the elements to avoid engaging in more than one at a time, which compounds the risks involved. Each element will be explained in great detail in each of the next five chapters.

5

Increase in Project Size

In Chapter 4, we briefly described the five common elements that lead to construction business failure. In this and the next several chapters, we will provide an in-depth description of each element, including case studies related to the element, and discussions of how to avoid or minimize the risks associated with these elements. This chapter concentrates on the first element, which is the most frequent cause of contractor failure: undertaking projects that are larger than a construction organization is accustomed to completing. Most contractors view larger projects as desirable. Some contractors feel pressured to take on larger projects for various reasons. There are, however, important reasons that a contractor should avoid taking on a project that is much bigger than the projects that have made that contractor successful.

Start-up companies have little choice but to grow for a number of years, eventually attaining a size that the founders are comfortable with or have targeted. Taking on larger and larger projects is one of the risks involved as start-up companies break into the industry. Having to repeatedly take on larger projects during a short period magnifies the start-up's risks. These growing pains are impossible to avoid and are a large part of the reason that start-up construction companies have an extremely high failure rate. Consequently, we did not include start-up businesses in our database. We only included businesses that had been operating for at least 10 years.

Established firms may choose to take on larger projects because of a lack of regular-size projects. The lack of their regular-size projects may result from a declining market, increased competition, local construction moratoriums, and other factors. In these situations, taking on larger projects may seem like a quick way out of the problem, and if the size increase is not too great, the risk is probably better than the alternative of running out of work. But as the project size increases beyond that of prior projects, so do the risks. A project that is 50% larger than any previous job carries risk, but much less risk than a project that is 100% larger.

Construction companies never run out of excuses for taking on a big project that eventually leads to failure: *"a good client wanted us to bid on it, the job was right next door to our office, we had an 'in' with the owner,"* and worst of all *"we adjusted our bid price after we learned some information about the other bidders' pricing."* Too often, bigger jobs seem to be too good an opportunity to turn down. In reality, research indicates that the risks are far too great for these projects to make good business sense.

5.1   Limits of Growth

It is not surprising that in this volume-driven industry, contractors regularly take on jobs that are larger than they have previously worked on, usually attempting to achieve rapid growth. For every business, regardless of the industry, there is a limit to the rate at which the business can grow safely. The challenge is to identify that limit before passing right through it. Organizational resources are stretched to the limit when increasing sales require a seemingly endless increase in people, financing, equipment, operations, and space.

Determining the limits of expansion is not easy because there are few rules or restrictions. In fact, some highly respected management specialists believe there is no limit to expansion. However, construction professionals should be hesitant about growing too quickly. There are so many companies in the industry that no longer exist, even though they were household names during their meteoric rise. Some people may point to specific reasons for each of these failures, but the reality is that rapid growth in and of itself is dangerous – not always fatal, but always a risk.

Because there are fundamental financial constraints to healthy and sustainable growth, growth must be carefully balanced with sales objectives, operating efficiency, and financial resources. Cases in which companies overreached for the sake of growth have filled the bankruptcy courts. The trick is to determine what sales growth rate is aligned with the realities of the company and the financial marketplace.

Every entrepreneur has limits regarding his or her abilities, available resources, and capital, and each organization is capable of doing only so much. During periods of rapid growth, construction companies change dramatically to the point that many would be correctly described as new, untested organizations. The prior smaller organization, which was so successful, is gone forever. If an organization is to grow, its management must grow.

Note that growing does not necessarily mean doing more of the same. Growing actually causes changes, and when people or companies change, they cannot easily return to what they were if things do not happen to work out. Growth is like progress, which moves only in one direction. There is no reverse. An organization's growth must be qualitative as well as quantitative. Qualitative organizational growth takes time and, to be effective, needs to occur prior to sales growth. Qualitative growth usually takes more time than it takes to capture larger projects. Almost universally, construction companies increase management size and skills only after additional work is on hand, not before. A common justification we often hear is *"once I have the work on the books, I can afford to hire that new project manager."* Growth for the sake of growth is risky in any business, but in the construction business, taking on projects larger than anything attempted before is always high risk.

5.2   Increased Risks with Larger Projects

The increased risks involved in larger projects stem from a lack of experience with the technical and administrative processes required for larger projects. An organization with a track record for completing projects of a certain size cannot assume it can make a profit on larger projects (Figure 5.1). Some have disagreed with this statement, saying, *"We used to do much smaller jobs than we do now, and we're still making money."* The question we would ask in response is *"How long did it take before you became*

Figure 5.1 The size and complexity of a project should be closely aligned with a contractor's profitable experience to effectively manage risk.

profitable after you began taking on bigger projects?" As already stated, growth of 15% or less annually has historically been safe growth. A company may be able to successfully double the size of its projects over time, if the growth is gradual. Additionally, even if the company has successfully doubled in size once, that does not mean the company can double in size again and again. A significant growth rate might sound impressive, but it has too often turned out to be disastrous. Growing successfully is a matter of risk management, and it is clearly not easy for contractors to know just how large a jump in project size they can take on while maintaining a modest risk. At a minimum, it is a prudent business practice to fully comprehend the nature of the risk and how to evaluate it before putting the business at stake by taking on a giant project.

Case Study: Apartment Complex Disaster

This case study focuses on a successful construction firm that had an average revenue of $16 million a year and produced commercial and apartment buildings. The company had been in business for 15 years and had experienced steady growth doing two or three major projects a year, and a lot of smaller work in a three-county area of one state. The average-size project had grown proportionate to the company's growth. The company considers anything near $3 million to be a major project, and the largest project to date was $3.5 million. One-third of the annual volume consisted of small jobs, many under $500,000. Profits were good, and development in the company's area was on the upswing.

An out-of-state developer announced plans to build a luxury condominium in the area, and the contractor's estimator requested the plans. After seeing the size of the project, the contractor almost sent the plans back. The project was enormous for the contractor, likely about $8 million, but as far as the contractor could tell, they were the only local company bidding for the job. The other bidders were larger contractors from out of the area. The company thought they had a real competitive advantage and decided to bid on the job. The contractor had three major projects underway at the time, but only one was bonded. Since the one project was almost complete, the company was able to get approval from its surety for the bid bond, but the surety remarked that the project seemed large for the company.

The design was first class all the way, and as the bid date neared, it became obvious that the project would be larger than the $8 million initially thought. The company was now guessing closer to $9 million. The contractor had some difficulty getting prices together because some of the specialty items were from distant suppliers and new sources. The size of the electrical and mechanical work for the project precluded the option of using the local subcontractors that the company was used to working with. Consequently, the estimator had to interact with strangers on some very sophisticated systems and controls. He was concerned about verifying that everything that was required was included, but not duplicated. There was a last-minute snag regarding the bonding, but a hastily arranged meeting overcame that issue, and the contractor submitted a final price of $9.5 million. The effort the contractor and his team had to put into estimating was so extensive that they had to pass up several worthwhile projects in their normal-size range. The contractor was awarded the project.

The contractor had taken into consideration the specification requirement of having a full-time project manager and a full-time field superintendent on the project. However, he had convinced himself he would not need the project manager on site for at least three months, until the project got rolling.

No sooner had the job started than the developer's full-time field representative insisted that both positions be on-site starting on the day work commenced. The company had three key field superintendents, and they were running the three other major projects the company was working on. The owner had always considered himself the project manager for all the work and had planned to put his best superintendent on the job as both project manager and superintendent. The developer would not agree to that arrangement, so the firm's best superintendent was assigned to be the project manager, and the second-best superintendent was assigned to be the project's field superintendent. Consequently, the least-experienced superintendent was left to run the other three major projects. Within days of the start of the new project, the project manager began to lay out the site for foundation excavation when he was reminded by the developer's representative that the specifications called for a licensed surveyor to lay out the project. The contractor began to see that this was going to be an entirely new game. His organization was totally unfamiliar with some of the construction processes required in the specifications.

Once the project was underway, both the contractor's owner and his estimator had to spend what seemed to them an inordinate amount of time attending site meetings, updating schedules, and reviewing shop drawings and submittals. The shop drawings were a particular problem because the company had never previously been required to formally approve every page submitted. The owner and estimator correctly reasoned that if they were going to approve all submittals and shop drawings, they needed to be reviewed carefully. It took a lot of time, and the contractor eventually hired a full-time draftsman to handle the preliminary review and coordinate the submissions.

The contractor had assumed that the developer would overlook some of the specifications he considered to be excessive, as other clients had in the past. This assumption turned out to be very costly. He had to hire an on-site project engineer even though he did not believe he needed one. He also had to follow the strict and expensive emergency and first-aid requirements laid out in the specifications. He had never worked on a project where the client had a full-time representative on-site. On previous projects, a client's representative would occasionally visit the site, and the visits were always pretty laid back.

The payments on the project were the only thing that was not by the book. Approvals were very slow, and the contractor became frustrated when he was unable to get through to anyone with authority at the developer's home office. Four months of payment requisitions had gone in, the fifth month's requisition was being prepared, and no payment had been received. Finally, after the contractor had maxed out his line of credit to finance the project and keep the job moving, he threatened to stop the job if he was not paid. This threat got the attention of the developer, and the contractor was invited to the developer's out-of-state home office.

Over an elaborate lunch, a senior member of the development firm (who happened to be an attorney) explained that the developer would pay the contractor in accordance with how the developer paid everyone, in the normal course of business. But the attorney made it clear that another threat to stop the job would also result in the contractor being removed from the project. Also, the developer was not happy with the project's progress, pointing out that the project was 13 days behind the schedule that the contractor had submitted and that had been included in the contract. The contractor was told to expedite the work to avoid back charges for any losses the developer might incur if the project was delivered late. The last thing the contractor was told was that his payment had already been mailed. The contractor went home, managed to arrange a line of credit that was a little higher and took out a second mortgage on his home in order to continue moving forward on the project. When the promised payment arrived, it was for the first month's work – a big surprise to the contractor.

Meanwhile, discontent on the project increased among the ranks. The project manager and field superintendent felt they were being overworked because this project was outside of their expertise and comfort zone. There was also a lot of overtime work trying to meet the schedule. As the job progressed, the level of activity and number of tradesmen involved became higher than the project manager or superintendent was used to or comfortable with. The constant presence of the owner's representative, who was now accompanied by two other inspectors, consumed a lot of the project manager's and superintendent's time. They were overwhelmed and were becoming very apprehensive. The cost reports on the project were not looking good, and the owner was pushing the project manager pretty hard about the schedule.

Eventually, the project manager decided to take a job at a different company. He was genuinely apologetic when he quit, but he explained that he simply could not handle the pressure. With so many job openings in the area, there was little chance of hiring qualified individuals from outside the company, so the contractor promoted the project superintendent to the role of project manager and promoted a good foreman to the role of superintendent. The contractor was surprised when the developer's representative approved the foreman's appointment as the superintendent. (What the contractor did not realize was that the developer's representative knew he could get more out of the contractor if the field management was not particularly strong.)

By the time the project was two-thirds completed, the contractor knew he had a substantial loss on his hands. He did not know if his bid was too low – the bids had not been opened publicly. (In fact, the bid was less than 1% lower than the second bid.) He was not sure if his bank would extend his line of credit further, and his cash flow problems were mounting. These financial problems were compounded by the fact that two of his other three major projects had finished poorly. He knew that the poor performance on these projects was the result of moving his best men from those jobs to the larger project. He had expected his least-experienced superintendent to complete and close out the three projects at the same time. The contractor had planned to help the overworked superintendent but had not been able to because the contractor had to spend all his time at meetings and solving problems on the big job.

At about this time, the contractor came to another realization. He was in the middle of a construction boom in the area but was running out of work. He was doing fewer small jobs than ever before because his estimator had not had time to bid on more of them. The head estimator was spending a huge amount of time managing the office because the boss was spending so much time managing the unanticipated administrative and paperwork requirements on the larger project. The smaller projects had always been profitable and were now sorely missed.

The larger job had caused other problems too. For example, just after the contractor was awarded for the big job, a very attractive job became available from one of the contractor's best clients. After spending a lot of effort putting together a bid, the contractor was shocked when he could not get a bond for the project. Though this new project was the contractor's average-size project, the contractor could not get a bid bond because all of the contractor's bond credit was tied up in the large project. The contractor needed more (profitable) work badly and was promised additional bonding capacity as soon as the contractor's year-end statement was available. One reason for the delay was the exceptionally high receivables from the big job, and the surety was getting nervous.

When the financial statements finally came out, they were not good. Payables had dried up most of the contractor's cash flow, and tight finances were causing all the work to lag behind schedule. Several subcontractors complained about nonpayment, and a couple had notified the bonding company of the payment problems. Not long afterward, the bank placed the contractor in default, causing his line of credit to vanish. When his funds ran out, he was forced out of business.

The bank's sudden line-of-credit reversal is not uncommon, but some readers may not be familiar with the reasons. The contractor, like most construction company owners, had signed a blanket loan guarantee with his bank years before to establish his line of credit. He signed both corporately and personally, and his wife also had to sign, which is not uncommon. By signing personally, the owner's personal assets, including the equity in his home, were part of the borrowing arrangement. As mentioned above, the contractor had taken out a second mortgage on his home to fund the big job. When the bank discovered the second mortgage, which was a violation of the loan agreement, they ceased lending.

Obviously, it cannot be said with certainty that the contractor would be in business today if he had not taken the big project. But given the favorable construction market, the contractor certainly did not need the big project. That fact makes the end result particularly poignant. The contractor was profitable and well-positioned in a good marketplace. It was reasonable for the contractor to think that being the only local bidder was a competitive advantage. However, because of the size of the project, should the contractor have even considered the project? Even if the project remained the $8 million initially estimated, such a large project was outside of the company's experience. Of course, the contractor and estimator should not have been surprised that the initial estimate would increase (to $9.5 million). They had no experience with projects that size and had also forgotten that their previously biggest project had increased from their initial projection of $2.9 million to $3.5 million.

This disastrous project was in the contractor's own backyard and involved the type of construction that company did best. But the project was much bigger than the customary size. In such a case, a contractor should not assume they will be able to foresee the impact of that project on other work, cash flow, bonding capacity, and profitability. Clearly, the contractor did not realize the tremendous risk involved in taking on a construction project that is substantially larger than anything the organization had previously completed. It is safe to say that if the contractor had recognized the risk, he would never have considered taking on the project. So many cases are similar to this one that by now it should be common knowledge that taking on significantly larger projects comes with huge risks. Success is partly based on defining a construction organization's niche in the industry, and that niche includes project size.

5.3 Underestimation of Project Size

Often a contractor's initial projection about the size of a large job they plan to bid on is underestimated. One reason is that when an organization bids on a project that is considerably larger than prior projects, the organization tends to relate the work required to the scale of the work they and the company are accustomed to. This tendency is particularly prevalent for work that is not estimated by unit price but by activities, such as layout, equipment setup, and cleanup. When relying on prior experience to estimate the hours required for activities, it is easy to forget that the project size affects the degree of difficulty and therefore the time required.

A person is more likely to make errors in estimates if the person is experienced in estimating small projects but is now estimating a large project. The same is true when a person with experience in estimating large projects is now estimating a small project. We are all prone to scale things down or up to conform to our previous experiences and expectations. In reality, the project size matters. For example, the equipment needed for a large project may be two or three times bigger than the equipment needed for a similar but smaller project, whether the work involves constructing a building, road, or bridge or completing plumbing, plastering, or painting. Repetitive work versus nonrepetitive work matters. High work versus low work matters. Shallow work versus deep work matters. Underestimating the effect of size is like thinking we can jump across a six-foot-wide trench because we can jump across a five-foot-wide trench. In a construction project, we cannot estimate the cost of placing 1,000 square feet of work by using the unit cost we incurred when placing 20,000 square feet of the same work elsewhere.

Case Study: Sewage Treatment Plant Failure

This case study focuses on a second-generation utility contractor that was regularly involved in renovation, expansion, equipment upgrades, and modernization of sewage treatment plants. The largest of these projects to date was $7 million. Following their decision to grow, the company was awarded a project similar to their experience except for the $10 million size. Six months later, they were awarded a $14 million project and shortly, thereafter the company bid on and won an $18 million job. There was great concern about this bid, because the next two bids were considerably higher at $22.5 million and $23 million.

Early in the $18 million project, the company encountered serious cash flow problem and their surety had to step in. The entire bid was extensively studied by the surety company's consultants, and there were no major mistakes. However, a detailed analysis of the estimate determined, to the amazement of the contractor, that almost all of the hundreds of line items were low. Although this project involved the same kind of work the contractor always did, no one who worked on the estimate, or anyone else at the company for that matter, had ever worked on a job this size. So, they apparently scaled the job down in their minds to coincide with their experience and expectations. (Psychologists refer to this phenomenon as successive approximation toward a goal.)

The surety's consultants noted that the scale in the drawings for the $18 million sewage treatment plant was smaller than the scale of any other projects the organization had ever bid on. The plan documents of course were the usual size, and the estimators had properly noted the correct scale in their quantity-takeoffs. The estimators had simply downsized the project in their mind and esti-mated too low in too many places. At this point, the bids on the $10 million and $14 million projects were reviewed, and they too were considerably low, but not as seriously low as the bid on the $18 million project.

Following the analysis, the contractor and estimators all agreed that the bid price was wrong. The losses on the $10 million, $14 million, and $18 million projects put this previously successful company out of business (Figure 5.2). Years of hard work and successful projects were not enough to save this company after it took on large projects that it should not have attempted. They were simply too big for this company, which had been in business for over 50 years.

The story and lessons do not end here. The bonded work, which in this case was all their work, had to be completed. Sureties have options for completing the projects of failed clients. They can support the insolvent company in completing the work or delegate the projects to other bonded contractors. Because the $18 million project involved highly specialized work that the organization was skilled at and the contractor was considered trustworthy, the surety decided to support the con-tractor in completing all the work. Through this process, the contractor was surprised to learn more about the estimating missteps that resulted from bidding on a significantly larger project than the company had experience with.

The contractor's estimate on the $18 million project underestimated the number of office, storage, and crew trailers by 100%. Twice as many were required to facilitate the work. The estimate included the use of two company-owned cranes and one rented crane, but four cranes were actually required, all of which were larger than the ones estimated. The company's cranes were not large enough, so all four cranes had to be rented.

The project bid also underestimated the number of workers required. Organizations with no experience in larger projects often believe they can run a large project with the same number of people as required for smaller jobs. In reality, organizations need to determine how many key people a job will tie up and for how long. Then, companies need to look at what other work they have and how the big project will affect the other work. The impact of tying up key people needs to be evaluated, as does the inability to go after

Figure 5.2 Successively larger-sized projects of the same type can result in significantly more risk.

additional normal-size work. Of course, if little other work is available, the inability to go after normal-size work is of less concern, but someone has to ask the hard question: *Do our key people have the right experience to handle work of this size and can we do this job and make a profit?* If additional tradespeople have to be hired for the project, the risk is increased because the abilities and loyalty of new people are untested. If the new hires do not work out and need to be replaced mid-project, another set of problems develops.

It is also important to understand that the client's representatives, designers, and inspectors will probably have more experience with the size of the project than the company's field managers. As seen in the apartment complex case study, the contractor's project manager was in over his head with the technical requirements of the project and didn't have the appropriate experience to effectively handle the project requirements, or to deal with the qualified owner's representative.

Case Study: A Road to Nowhere

This case study focuses on the largest road builder in its area. Located in a midsize city in a lightly populated state, the contractor built and repaired city and county roads, and a large number of parking lots and commercial driveways. As a successful second-generation family business, the contractor had progressed through the high-risk leadership-succession process. The company was approaching 60 years in business when a very large interstate-widening project came up for bid in the state. The company had never attempted a federal project before, but the project was certainly the company's normal type of work and the company decided to go after it (Figure 5.3). The bid was considerably low, but the contractor rationalized that the reason was the contractor's superior local knowledge, as the other bidders were from out-of-state.

Figure 5.3 A federal highway project can be much more complex and impose significantly different risks to a contractor compared to a municipal roadway project.

The surety after providing the bid bond expressed concern about the low price, but the contractor insisted it knew what it was doing, and the payment and performance bonds were issued. During the first weeks of the project, the contractor began complaining that the inspection requirements were interfering with production. For example, while the contractor was pouring a small manhole foundation, the inspectors insisted on getting three concrete test samples before the work could proceed. In addition, the inspectors took the temperature of the concrete as it came out of the trucks and on numerous occasions sent the trucks away half full, which delayed progress. The contractor had never experienced inspection with this degree of scrutiny before.

When winter set in the inspectors stopped the work because the specifications did not permit concrete or paving work to be completed when temperatures were below freezing. As a result, none of the schedule milestones for the first season of this two-year project were met, and the contractor and surety were warned in writing of a potential default if the schedule was not achieved. Notified of the risk of default, the surety discovered that the contractor knew that they would not likely be able to complete the project during the next season. The surety put a different contractor on the project and discontinued bonding with the now-defaulted contractor. Because most of the contractors' typical projects were bonded, the inability to bond work effectively put the contractor out of business.

It is easy to think that what happened is unfair, but be reminded that business is business and not necessarily fair when specifications are not met. Although this case includes a change in the type of work, the case is also an example of the risk incurred with a change in project size because the contractor would have survived if it had taken on a small federal project instead of a project that was among the largest the company had ever attempted. (*Why? The company may still have lost money on the project, but the company would have been more likely to survive a smaller loss.*) And the company would have learned a valuable lesson. After the learning lesson, the company could have decided not to pursue federal work again. Or the company could have decided that if it pursued federal projects in the future, it would add to the project estimate the cost and schedule impacts of the inspection procedures and specification requirements in federal contracts.

It is easy to want to blame others for this business failure, but the skills and knowledge required to be and remain a successful contractor include a full and detailed understanding of the common elements of construction business failure.

5.4   Clients and Retainage

The risk of undertaking a larger project increases when the client is new. It is always important to know something about the client, but if a company is out of work and desperate, the company will probably attempt to take on a larger project even if unfamiliar with the client. Among the information a contractor should learn about a new client is its payment procedures and reputation. Understanding the payment procedures is particularly important because they will affect the contractor's cash flow.

Contractors should take a realistic, if not pessimistic, look at the length of time the job will take and ensure that they have accurately planned for retainage amounts. If retainage will be reduced when the job is 50% complete, it is essential to determine whether this reduction is the client's option and if it depends on client satisfaction. Remember that many clients consider it prudent to hold back retainage, and if there is any question at all about whether to hold back retainage, it is much easier not to pay than to pay. Therefore, contractors need to determine the effect on their cash flow if the retainage is not reduced.

If retainage is payable only after final acceptance, it is essential that contractors take a hard look at how long it will realistically take to collect when dealing with a new client. This planning should take place before the bid is submitted, not after the project is awarded. Well-managed companies predict how much cash they will need at various times in the future and then attempt to raise that amount before it is required. At the very least, they make plans for securing the cash when they need it. For long-duration projects, it is good practice to conduct a cash flow analysis on a project to develop a thorough understanding of the working capital requirements the project will impose on the company during the project. The results of this cash flow analysis should be considered in the estimate when bidding on the work. Huge unpaid retainages are common among distressed construction enterprises that find themselves at the mercy of their bank lenders or sureties.

Once a contractor undertakes a larger project, it is critical to give it the time it deserves, as it will represent a significant portion of the total volume. However, top managers need to look carefully at the allocation of time and not forget about the other work on hand. The other projects may be smaller, but they will continue to be the company's bread and butter. An organization seldom wants to risk the profits that small jobs generate, and it may not be able to afford the losses that could result from a lack of attention on these projects.

5.5 Alternatives to Taking on Large Projects

If a contractor is running out of work because of a declining market, what are the options other than taking on larger projects? The hardest option to discuss in the construction industry is to do less work. Most contractors will not accept such a notion, but it is a very viable alternative. Cutting back on overhead and becoming a smaller business that fits a declining marketplace is a very realistic approach. You might even describe it as *cooperating with the market*. If the entire market in an area is soft, then all of the contractors will be looking for work. Larger contractors, who do not usually compete with midsize firms, will go after the midsize work. Mid-size contractors are often forced to pursue smaller work. And small contractors have nowhere to go.

Bidding on a normal-size or smaller project is far less risky than bidding on larger projects. The problem, of course, is that an organization's annual volume will necessarily drop when taking on smaller projects because, in a soft market, there are fewer jobs and increased competition. An organization can strive to get only a share of jobs that will produce a profit or at least allow the organization to break even.

Another alternative is to expand the company's work area to find projects that are the company's normal-size and in the company's niche. The risk involved in geographic change is discussed in a subsequent chapter. Here, we will only say that this option can be explored and balanced against the risk of undertaking a very large project. Unfortunately, most construction businesses are not very flexible. They are not set up to expand and contact with the availability of work. A good example of flexibility is a general contractor that was originally a concrete contractor. Every time work slowed down, the contractor would take on a few concrete subcontracts to keep busy.

A construction organization should have flexible overhead and be cautious of building a large organization with fixed overhead in a volatile marketplace. Approximately 15–25% of overhead can be flexible and easy to shed, such as short-term leases on some equipment and some temporary personnel in administrative positions. Instead of hiring permanent office staff each time a company expands, it can use the services of temporary placement services until the organization expands again. The same can be done with vehicles and equipment fleets, where a portion can consist of short-term rentals, allowing much greater overhead flexibility.

Permanent overhead, which cannot be easily reduced, is what requires construction companies to desperately compete for sales in a declining market. Submitting bids that allow for little or no profit just to support overhead magnifies risk and in a declining market can be suicide. Some construction executives are convinced that they cannot let go of overhead – particularly their people and equipment – because it will be needed when the business cycle turns around. This argument is compelling, but if the company does not survive a down market, the argument is irrelevant.

Business is about recognizing, minimizing, and managing risk, not about maintaining size or potential in a future market. Every construction enterprise must operate in its current market by using past experience to navigate. Large construction projects have substantial risk and are to be avoided if the company lacks experience.

There will be occasions when, for whatever reason, an organization will decide to undertake a job much larger than anything they have ever done. Hopefully, it will consider all of the alternatives, and weigh the risks involved, and if the company decides to move forward, it should develop a course of action and stick to the plan.

5.6 Summary

We have spent considerable time in this chapter cautioning against taking on projects larger than an organization has experience with. Our focus has been quite deliberate because the risks are so great that they should be avoided if at all possible. If a company takes all the precautions suggested here, there is still no guarantee it will succeed when taking on a larger project. Do not believe that constructing five one-story buildings is anything like constructing one five-story building, any more than completing five $1 million jobs is like completing one $5 million job. Experience in one project size does not prepare an organization for similar projects larger in size. An organization must learn how to crawl, then how to walk, then how to run, and finally how to fly. Leave out a step, and the company may fall from a considerable height.

6

Changes in Geographic Area

Changing the geographic area a construction business works in is a significant change and far from business as usual. The risk in moving from working in a known geographic area to an unknown area is very real and can be evaluated and measured. If a contractor must make this change, a prudent amount of caution and planning is necessary to minimize the risks.

6.1 Business in Your Normal Area

Business as usual for a construction company involves profitably bidding on, winning, and completing work in a specific geographic area at a profit. We refer to this as a firm's *normal area*. Doing work in the normal area never guarantees profit on a project, but there is a greater likelihood that profit will be produced in the normal area because the organization has profited there in the past. Continuing to do business in the area involves only modest risk (in this industry, no project is risk-free). The risk is significantly higher when taking on the first job in a new area.

The normal area is not necessarily an area of 10 or 100 square miles but is defined as an area of success, and it may be irregularly shaped. For a firm that does remodeling in the suburbs across the river from a big city, a job that is a stone's throw over the river in the city may be out of the firm's normal area. Contractors need to know what their areas are and are not. The size and shape of the area varies from one company to another, but the premises remain the same. Leaving the typical work area to pursue a project elsewhere involves working in a place where the organization has no experience operating at a profit. We are not suggesting that the job will not get done, but we are noting that the ability to make a profit is in question. Even if a profit is generated from an out-of-area job, there are greater risks involved than when working in a known territory.

6.2 Reasons for Changing Geographic Area

There are many good reasons for expanding a construction business's geographic reach, including a desire to grow, a shrinking local market, and the opportunity to follow customers or designers to new areas. Even if the reasons are very logical, the company needs to measure the associated risks

Figure 6.1 Local working conditions can be unfamiliar and difficult to accurately predict when taking on work in a new area. *Source:* Vasiliy Ulyanov/Adobe Stock Photos.

and carefully consider whether branching out geographically is wise (Figure 6.1). Numerous contractors do not give a second thought to jumping vast distances, sometimes two or three states, to get work. Some expand geographically as if doing so were the most natural thing in the world. However, many that have made this leap have reaped poor or even disastrous results, particularly if unprepared organizationally. Often, contractors that are busy developing or expanding their market do not have the time or even recognize the need to develop, expand, or improve their operational systems and capabilities. As a firm increases in size, the organization and its systems are strained and usually need to be modified to cope with the growth. And, of course, the adjustment should be taken care of prior to the growth occurring.

Case Study: Project in a New Area

Consider the following case study: A utility contractor was asked to bid on an underground sewage piping project 350 miles away from the firm's normal area. Their normal area was the area within a 100-mile radius from the office. The sewage piping project was in a remote area, with only two other contractors bidding. The design engineers, well-known to the case study contractor, asked the contractor to participate. The contractor did not need the job, and the project was at the high end of the project size the firm typically worked on, but the fact that the design engineers invited the contractor to bid and that there were only two other bidders made the project tempting. Although the firm had never worked on a job so far away from the office and from supply and service providers, the firm thought it could keep the project fairly self-contained so that it would not drain resources from other projects. Further, the contractor thought that it could try some special expediting techniques to shorten the overall project duration, thereby limiting expenses. Winning the project would provide an increase to the revenue and profit the company expected to make that year, but the contractor did not really care whether he got the project or not.

The firm proceeded to estimate the job in the usual manner and even arranged for one of the project managers to take a day trip to the site. He did not see anything unusual about the site's conditions. In fact, everyone thought that the remoteness of this farming town would mean fewer traffic problems resulting from all the street closings that would be necessary. At the local coffee shop, the project manager asked about labor and was told that a lot of locals were out of work and that the firm could get 30–40 people the next day if they wanted them. The project manager was also told that the firm could get quite favorable rates at a local motel if several rooms were reserved for an extended period. The contractor priced out the job as if it were in the firm's backyard and then added costs for equipment transportation, mobilization, and housing of key people. The firm even added a markup for the nuisance factor and got the job.

The specifications called for the project to start 14 days after the project was awarded. Starting then would put a strain on available equipment, but the firm could lease a couple of pieces in its normal area to replace what was sent to the remote job site. Also, there appeared to be some available farm equipment in the area that could be used for grading and seeding, and there appeared to be trucks available locally. However, immediately after the firm mobilized, it had problems obtaining county permits, which the firm had assumed was a one-day process. One of the senior engineers had to travel to the location three times to assist the friendly design engineers to help solve the issues. The problem was that the county engineer had no experience with projects of this size. Ultimately, because of issues in getting the permits, the job did not get going again for three months.

The contractor had figured that the project would take four months from start to finish, but while waiting on the permits, they decided to cut the time in half by having workers and equipment operating 12 hours a day, 7 days a week, with every other weekend off. The workers would earn a lot of overtime, and the firm would save on housing costs because of the shorter project duration. Since the firm wanted to finish the job in just two months, the late start did not seem like a problem.

After the permits were obtained, everything started out as planned, except that some lowboy and tag-along trailers were unavailable to use for local projects because they were being used to move more equipment than originally planned to the remote site. If the remote project had started as originally scheduled, the large number of trailers would not have been needed at the same time. Because the firm thought there was plenty of extra money in the bid and wanted to maintain good relations with the designers, the firm agreed not to pursue additional compensation from the owner for the delay caused by the permit issues.

As the work progressed, the distance began to take its toll. It took a day or two to get necessary items to the site. The downtime resulting from equipment maintenance and obtaining repair parts and required supplies became a nightmare. Also, the firm experienced a big surprise when it became apparent that the crews and equipment were very unwelcome in the town. The project had been planned by a town administration that had been replaced in the recent election by individuals who ran for office on a promise to stop the project.

Figure 6.2 Contractors may not always be welcome to a new area, which can result in unnecessary and expensive requirements being imposed on the newcomers. *Source:* Susan Vineyard/Adobe Stock Photos.

The conservative townspeople did not want the sewage collection system, with its attendant "*costs and mess*," as they put it. They had used septic systems for decades and were content to continue using those systems. The part-time mayor read the specifications from cover to cover and held the contractor's feet to the fire. Street openings had to be protected with flashing lights and flagmen, even on dirt roads. The mayor further insisted that the lights and flagmen be provided until all paving repairs had been completed and accepted. The flagmen were needed four times longer than anticipated for the firm to comply with the letter of the contract and protect streets that had no more traffic than about 20 vehicles a day (Figure 6.2).

Many of the details are too painful to present here. When issues on the project arose, the design engineers were very cooperative with the contractor early on in the project, until the town administration threatened to throw the engineers off the job. From that point on, there was no salvaging the project. The design engineers also wanted overtime pay for all the inspection time caused by the contractor's 12-hour-a-day and weekend schedule. Because of the difficulty and delays involved in getting supplies, the firm began ordering more supplies than were needed, to make sure the team did not run out of anything. Material costs on the project ran over the estimate by almost 20%.

The project started in mid-September, and when the harvest season began in October, all local labor left the project. Some of the workers were replaced by transferring employees from jobs in the normal area, but there were not enough workers to maintain the current schedule. The specifications stated that no seeding could be done after 15th of October and no paving could be done after 30th of November. Because the firm did not meet these deadlines, the contractor had to pause the work until the spring. Ultimately, labor costs ran over by almost 200%. The total loss on the project exceeded 60% of the total contract price. The final payment took two-and-a-half years to collect, but by that time the firm was no longer in business.

6.2.1 Adding Insult to Injury

If the far-away project had been the only loss during the year, the contractor might have survived the experience. However, the remote job affected the home-area work in unexpected ways, creating additional losses. The contractor's plan was to use their best employees on the remote job. The project manager was handling three other contracts close to home when he was sent to the remote site. The three other projects were assigned to other project managers, and the changes in field management resulted in two of those projects losing money. Similarly, the two best superintendents were transferred to the remote job, requiring that other projects be assigned new superintendents mid-project. When the remote project got in trouble, a third superintendent was transferred there, further disrupting the other projects.

On short notice, the company's five best equipment operators were taken off the local projects they had worked on and were transferred to the remote project. It is difficult to measure the impact that this shift had. Further, key mechanics, many of the service trucks, and even the delivery men were not available in the home territory at various times during the remote project. Productivity and morale throughout the organization were worse during this period than at any other time in its history. The overall result of taking on the remote project was that six tradespeople quit and seven jobs lost money. Before the firm bid for the remote project, the company was four months into the fiscal year and enjoying its usual profit margin. If the loss from the remote job was excluded from the year-end figures, the firm would still be in the hole because of the losses on the other projects. Including the loss from the remote project, the year-end results showed a devastating loss.

We do not intend to imply that such a disaster will result every time a contractor changes its geographic area of operations. What we are indicating is that there is a huge risk in moving into unfamiliar territory. Why is the risk huge? Because even a miniscule risk of bankruptcy is huge (not unlike the risk in Russian roulette). Was the remote project a preventable problem? Without a doubt. For one thing, the contractor did not care if it got the job. So why go after it? It seemed like a good idea at the time. No one in the organization knew about the published research that confirmed the risks involved in expanding a firm's geographic area. Had the firm understood the potential risks being undertaken, the firm would have at the very least checked out the project location and local conditions more carefully. Risk-takers need to know how much risk is involved – minor or major, a little or a lot?

Should contractors check the political climate everywhere they go? Probably. When contractors work in their own backyards, no matter how big those backyards are, they usually understand the political climate and are likely to encounter fewer surprises. Contractors also need to consider the costs associated with the greater distance of a project. Transportation is expensive, but at least it can be calculated. The impact of distance on performance and production is almost impossible to calculate if you haven't experienced it. In the construction industry, contractors are constantly challenged with the question: Is this project worth a shot? *A shot* is the perfect choice of words because it is usually *a shot in the dark*. Your call.

The important point is that this remote job had to be performed differently than the contractor's other work. Distant jobs are different and may need to be handled differently. Problems arise when a firm does not realize this fact. If an organization is working in a new area, the organization should assume it will encounter different conditions. In the case study, the contractor's ability to make a profit was drastically reduced by underestimating the disruptions and costs resulting from moving equipment and workers to and from the job. Contractors must not underestimate the loss of efficiency that results from the inability to quickly and efficiently maintain equipment. The firm had not realized that its ability to efficiently move employees, equipment, and supplies in its normal working area was a major factor in its profitability. The distant job presented a totally new experience.

Some actions can be taken to minimize and control the risks of doing work outside your normal area. However, these actions are not as important as thinking hard about a project before submitting a bid. Plenty of contractors have completed work outside of their normal areas and made money. Some have even opened regional offices and have been successful. But even the contractors who succeeded were at enormous risk, whether or not they recognized it at the time.

6.3 Risk Management in Long-distance Projects

The safest way to expand beyond the normal work area is to begin at the edge of the existing work area, with a typical-size project. To manage risk, the smaller the project the better. A company should test the profitability as the first project progresses and attempt to discover if there are any limiting factors. Only after the test project is complete and is a success should the company try another distant project. If the project is a success and the company wants to continue expanding, perhaps it could move along an interstate corridor, along one side of a river, or up to a state line. Expanding this way limits the changes and risks that may await a contractor in a more distant location. It is easier to pull back when expanding from the perimeter than from a remote location. Dealing with recalls, guarantees, and required maintenance periods is a lot less expensive at the perimeter of a company's normal area than at greater distances. A company can continue expanding in this manner unless profits begin to drop off. Even if the company needs to pull back, it might be able to reenter the area after some thorough planning.

It makes sense to assign trusted employees to a project in a new area, but doing so is costly and reduces resources for normal-area projects. Hiring locally is also an option but will require training and orientation in order for the work to be performed the way the company has always managed its work. It is important to recognize that this training and orientation will add to the time and costs associated with the project. Expansion has a greater probability of success if a firm brings superior procedures, proprietary processes, and/or better skills into an area. If a company finds out that things are just not done in the manner they are accustomed to in the new area, fitting in will be more difficult, and the company may want to reconsider whether they belong there.

Although most of the workers may be local, relocating one or more top field people to a remote project is a common and appropriate approach. On the positive side, the organization is likely to get consistency in methods used and honesty in reporting. On the negative side, relocating employees reduces home-territory resources. A small or midsize contractor that wants to expand geographically should fully understand that one of the major risks of growth is that it stretches existing resources.

Initiating geographic expansion should include taking on one and only one project at a time, and the job should be in the middle or the lower range of the company's normal project size. This approach should be considered a mandatory method for managing risk, whether the new location is near the perimeter of the normal area or at a more distant location. Too many contractors take the position that they are not willing to travel a great distance for a small project, and, therefore, out-of-area projects have to be big jobs. That approach greatly increases risk. Another huge risk comes from taking on a second long-distance project before the results of the first distant job are in. Doing so makes no sense but is not uncommon because firms get caught up in the excitement of growth and expectations of positive results.

Because a distant job has a greater risk than a local job does, the distant job should be watched closely (Figure 6.3). The firm needs to regularly take the pulse of the job and accurately account for the use of company equipment. Company equipment is not free. Additionally, the firm should carefully measure the impact on home-area productivity and profit and attempt to quantify how the extra effort and attention that top managers direct toward the distant project affects the organization and its overall performance. Research shows that expanding too far is a mistake unless a senior leader is committed to being personally responsible for the results. If no senior leader can afford to devote the time and effort that expansion requires, then the expansion should not be attempted. There are very real constraints associated with geographic expansion.

Figure 6.3 Geographic expansion entails more than simply mobilizing equipment a further distance. Minimize risk by undertaking one job at a time and manage it closely. *Source:* MaxSafaniuk/Adobe Stock Photos.

6.4   Regional Offices

Another method of expanding geographically is to open a branch or regional office. This method has all the risks of enlarging the area normally worked in, plus the sunk costs of establishing a new office. Entering an established market means taking work away from others. One argument contractors make is "*the market is growing, so we are not competing for local contractors existing work. There is going to be plenty of work from local contractors and for everyone.*" Though some people think this argument makes sense, if a newcomer takes any portion of the growth in an existing market, it does not change the reality that the local contractors always consider that the market growth belongs to them. The local contractors will have smaller market shares of their market growth. Research demonstrates that increased competition, even in a growing market, always lowers the selling price, both for the newcomer and for the locals. This fact has been verified so many times around the country over the last several decades that it should not require further explanation.

The best way to address this subject is to clarify the typical dynamics of opening a regional office. To do so, we present the following case study.

Case Study: Regional Office

A midsize contractor had been in business for over 20 years when they were awarded a contract to build a local shopping center. The contractor had never built a shopping center before, and at $6.5 million, the project was among the larger projects the firm normally worked on. The firm put a lot of time and energy into the bid and then into the job and came out with a fair profit. During the next two years, the same customer awarded the firm two more shopping center projects. The firm's normal work area was half of the state. One of the projects was beyond the edge of that area, and the other was in an adjacent state. The in-state project was $8 million, and the out-of-state project was a little over $9 million. The contractor's plan was that the in-state project would be managed by one of the firm's senior project managers. The out-of-state project would be managed by a new local hire who was highly recommended and lived near the project site.

When completed, the profits on both shopping centers were a disappointment. The out-of-state project experienced some problems with rocky, difficult ground conditions early on, and both jobs were plagued by weather problems. The contractor was able to capture most of the tenant improvement work for the out-of-state shopping center, and that work made up for the loss on the shopping center. No such luck on the other project. The contractor thought that the distance to both project sites was the primary problem and that if they had an office closer to the sites, the projects would have been profitable.

Because the contractor became familiar with the area around the out-of-state site and because the tenant improvement work was profitable, the contractor decided to lease a small space in the shopping center and open a branch office. The project manager who worked on the out-of-state job was promoted to run the new office. One of the home-office estimators was willing to relocate to the new office, partially because he loved to fish in the area. Right off the bat, they were able to get a few small jobs in the area, and the new office appeared to be a great idea.

However, during the first year, the regional office managed to capture only $400,000 in work, not even covering the office's overhead costs. The difficulty in capturing work seemed to stem from the company being an unknown commodity in the area. The firm also learned that local competitors were taking every opportunity to spread rumors, some of which were false, about the newcomer having lost money so far. The new office was given one more year to become profitable. In the second year, the branch increased its sales to $4 million. Unfortunately, project profits still did not cover overhead costs. The office was still in jeopardy after the third year. It did not survive through the fourth year.

6.4.1 Getting Spread Too Thin

The first large shopping center the company built was $6.5 million, which was near the top size project the company had experience with. As such, the contractor himself gave that project a lot of personal attention. The job was in the organization's backyard and made good money for the company. The next two shopping centers were larger and farther away than the first shopping center. The only thing the jobs had in common was the client. The contractor had a good experience with the client on the first project and assumed the same would be true with the next two projects.

It was clear from the beginning that the second and third projects would not get as much of the contractor's attention as the first project did because they were farther away, and they would be constructed at the same time. Apparently, no one in the company, not even the contractor, ever correlated his personal attention to the first shopping center with its profitability and success. Getting the two big jobs at the same time caused celebration, not contemplation, and certainly not pessimism, in the organization.

A new project manager who lived near the job was originally hired for the out-of-state project in order to keep expenses down and to deal with local conditions and the local subcontractors. The project manager had been highly recommended by knowledgeable people in the area and did a decent job on the difficult project, but it lost money. This project manager, who became the branch manager, was untested by the company because of the lack of previous experience with the organization and lost money on his first project. The rock and weather problems could have happened to anyone, of course, but no one even wondered later about how he had handled the problems or whether the local bidders had anything in their bids to address these issues. The reality was that the new, untested project manager lost money on the first project he ran for the company and was promoted to establish a branch office.

The new branch manager had done some contracting previously and said he could bring in plenty of work. He brought in some small jobs but not nearly enough. No one considered the fact that the new branch office did not have any key personnel to do the smaller jobs, so the branch manager was charged with bringing in the work and also doing it. As it turns out, he did not do either activity very well. Likewise, no one thought it was imprudent for him to be distracted by the complex tenant improvement projects that were part of the motivation to branch out. The truth of the matter is that everyone in management was thinking of opening a regional office from the moment they got the out-of-state job, so there was never a serious analysis of the decision.

Even when it was clear that the two big jobs lost money, the firm quickly made the excuse that the tenant work would create a breakeven position for the entire endeavor, so it did not matter. No one acknowledged that home-office overhead costs associated with the big projects were not charged to the jobs and that neither were the home-office overhead costs of the tenant improvement estimates. The branch office's initial overhead, such as estimating, payroll services, phone bills, overnight delivery charges, and the new communication system, were also paid by home office. That caused the costs to date to be understated because much of the overhead costs were absorbed by the home office, which meant that the decision to open the regional office was based on incorrect information and assumptions.

Putting a trusted estimator at the new office was an appropriate idea, but the estimator used the same unit costs and overall approach he had learned at the home office. Doing so was a mistake because it was a start-up operation with no local cost history, no market analysis, and no forecasts. (Without any of this information, the difficulty in capturing more work should not have been a surprise.) Instead of starting slowly, gaining experience, and local knowledge over time, the contractor simply planned for the same rate of growth that the home office was experiencing. The growth plan started with a base of $9 million, the value of the original shopping center project, so it is no wonder that the $400,000 in sales the first year was a disappointment and did not come close to covering the branch office's overhead.

It is impossible to determine how the threat of closing the office after the bad year affected the estimator's actions. The threat might have motivated the estimator to become more aggressive in pricing the work and to become more competitive in capturing work. Indeed, the branch enjoyed exceptional growth in sales that year, but not in profits. The distractions and losses in the following year resulted from the work captured in the prior year. Consequently, the office closed.

The magnitude of the out-of-state losses might not have put the entire company out of business if the home office had earned its usual profit percentage. However, during the years the company built the second and third shopping centers and then opened a regional office, home-office operation suffered because of a lack of management attention, exceptional growth, and reduced resources because management's time was spread too widely and too thin. The result was that home-office profits diminished steadily over the next four years. As the performance of the big jobs and then the branch office deteriorated more and more, home-office resources and senior leaders' attention were directed toward the branch office problems. In this case, the failed regional office took the home office down with it.

Situations similar to the case study occur so often that it is almost an embarrassment to the industry. In this case, the contractor lost sight of how much his personal attention and effort contributed to the success of his organization. He forgot how much energy, night work, and worry went into developing a successful enterprise and how much his personal efforts contributed to the continuing success. Adding to this error, he put an untested person with limited support in a regional office and expected immediate profitable results.

A regional office is like a separate construction company. A home office cannot simply provide a new office with materials, money, and workers and expect the new office to somehow blossom into a successful construction company. A successful regional office also needs a leader endowed with the characteristics of business sense, natural instinct, drive, and a sense of timing. The leader must have the attributes of a founder and a coach in order to provide direction, support, and motivation. Project management experience does not automatically qualify someone to be a branch manager.

6.5 Establishing a Regional Office

Opening a regional office is almost a necessity if geographic expansion extends beyond practical travel distances. But opening a new office is extremely risky not just the first time but the first several times. The risk is even greater when a regional office is opened for opportunistic reasons, such as because of one big job in an area or because the firm is trying to attract a potential customer in the area. Opportunistic

Figure 6.4 Expanding into new areas can provide rapid sales growth, but experience in an area is necessary to ensure the growth can be accomplished for a profit. *Source:* Evgen/Adobe Stock Photos.

expansion, while highly motivated, is risky because it involves less time for planning, testing the market, testing the decision, and evaluating alternative locations (Figure 6.4).

Some may argue that large construction companies open regional offices all the time, so how hard can it be? The reality is that opening a regional office requires a specific set of skills that are not learned by building things. Because a lifetime in construction offers no exposure to the necessary skills to develop a successful business model, a contractor that opens a branch office is attempting something the contractor knows nothing about. The national and international construction firms that regularly open branch offices have gained the required skillset over decades, and most of these companies have specialists whose primary job is to research markets, plan locations, and facilitate the opening of branches. A firm opening its first branch office can only guess how to do it, which imposes a huge risk on the company. The reason the risk is so big is that there is no playbook or publicly accessible body of knowledge on how to open a regional construction operation. This trial-and-error method has bankrupted a number of previously successful construction companies while attempting their first or second branches.

Some of the basic elements are known. For example, the most important key to the success of a branch office is finding the right person to run it. The ideal, though many would say impractical, solution would be for the contractor to relocate to the branch office. Of course, doing so would only be wise if the home office is running smoothly and profitably. Otherwise, opening a regional office is suicide.

We recognize that having the contractor move to a branch office is impractical for a lot of reasons, the most important of which is that the home office requires his or her attention. Some would have you believe that establishing a branch office with just a good construction person and administrator will work fine based on the assumption that the remote office will just be an extension of the home office and will operate effortlessly. That model has never worked and has zero potential for success because the underlying assumption is in error. In reality, it is essential to select a person with the characteristics, knowledge, and talents of a leader (contractor) and to provide appropriate and initially (for years) extensive home-office support.

These are the minimum requirements for a prudent start-up operation, which is exactly what a branch office is – a start-up. We all know that the failure rate for new construction entities is frighteningly high, and the risks are just as real for new branch offices. One advantage of a branch office being the offshoot of a successful company is that the new office may not need to go through as long a growth and development period as the original did. The reason is mostly that the branch office inherits the reputation and perceived experience of the original company. However, these perceptions must soon be backed by actions if the office is to be successful.

If a new branch office is treated as a start-up and receives enough support from the home office, the branch may be able to advance without the missteps and mistakes that the parent company made as a start-up. Prudent risk management includes starting small, testing abilities and decisions, and growing cautiously. Starting out too big, even with the right ingredients, increases the potential for missteps and amplifies the already high risk. A modest start that spans a reasonable period of time is part of standard risk management for any endeavor and is particularly important for start-ups.

Treating the regional office like a start-up construction company and hiring talented employees are critical steps, but they are not enough. It is also essential to obtain local knowledge. And doing so should never be underestimated. As stated earlier, construction work is performed differently in different areas of the country. How different? In some cases, just a little different, but when profit is limited, it does not take much for a start-up to end up in the loss column. This fact is hardly ever recognized in first-time efforts to expand to new regions, because most people assume that things work the same everywhere. Sometimes the differences result from distinct local conditions, sometimes the differences result from local traditions, or both.

Conducting as much research as possible is a prudent step, but an organization that is new to an area primarily gains knowledge about the area by bidding on, capturing, and completing work. Therefore, the sensible approach is to do some work but not so much that lessons learned are too expensive. Through applying this approach, the regional and home offices get a feel for the new area and are able to make appropriate adjustments while the stakes are low. Unfortunately, most contractors think that they need to hit the ground running, so they get into the new market as quickly as they can. Suffice to say that this perspective is not part of a prudent risk management approach.

One of the common risks involved in opening a branch office is underestimating the costs involved. In too many cases, the costs are underestimated by a considerable margin. After a branch has failed, we often have heard contractors state, "*If I had known what this branch office was going to cost, I never would have opened it.*" Invariably, the costs are greater than expected and the income is slower to develop than anticipated. Unless an organization has considerable past experience of opening offices, there is nothing to base the estimated future costs and income on. Organizations without this experience would not know, for example, that financial contingencies need to be applied liberally to all projected costs and against all projected income. Doing so is the only way to compensate for the lack of experience and knowledge about opening a regional office. Minimizing the costs up front is critical to allowing for a withdrawal at less loss.

Overoptimism is one of the biggest threats because shortfalls must be paid for by home office and, in theory, paid back by the branch office. The potential costs and risks must be understood in advance so the company can be certain it can afford to open the office and, if appropriate, can discontinue the effort before it is too late. Unfortunately, what typically happens is that once a firm decides to expand, the excitement drives the expansion, and anyone who raises concerns is considered a pessimist. Some expansions have run out of cash not because of unexpected circumstances but because the planners overcommitted the organization.

Opening a branch office where a market is growing is a little less risky than opening a branch office for opportunistic reasons. A growing market or a need for the firm's specialty at least implies the availability of opportunity. The distance from the home office is important in locating a regional office when the motive for expansion is opportunistic, such as for a single *anchor* project. When the motive is a single project, the closer to the home office the better. When the motive is a growing market, the location is less important than the likelihood of immediate work. There is competition in every market, but breaking into a soft market is more difficult than breaking into a strong market, where there is at least perceived room for another contractor.

6.6 Contingency Plan

The final and exceedingly important risk management task involved in opening a branch office is to create a withdrawal plan. Even the most extraordinarily planned and executed opening of a branch office can fail. New operations seem to achieve inertia: once opened and set in motion, they continue in motion because everyone involved in the decision is convinced it will work. (*Failure is not considered an option.*) Sometimes the best business decision with the best plan in the world simply does not work out, which is why every plan should include withdrawal triggers and then detailed steps for withdrawing.

In creating a withdrawal plan, it is critical for the firm's managers to determine the parameters the firm will use to measure the performance of the branch office. It is wise to avoid too much optimism in expectations and to allow a grace period if goals are not accomplished by the specified dates. However, during planning it is imperative to establish a date and an amount of loss that will cause the firm to declare the branch office a failure and to close the office. Establishing a date and amount of loss is extremely difficult for planners because doing so seems incongruous with planning for success. It is essential to develop the withdrawal plan before steps are taken to open the branch, because once the effort is initiated, it is almost impossible to create a withdrawal plan.

When developing parameters for success for a branch office, financial measurements are most commonly used. Anyone planning to open a branch office needs to understand that it is close to impossible to recoup the start-up costs through profits during the first, second, and sometimes even more years. Though with luck the office will break even in the third year. Therefore, opening a branch office is a negative cash flow activity, which economists define as an investment. If the effort fails however, the *investment* becomes a loss for the home office.

The home office must determine in advance how much the cash outflow (loss) will be. Both best-case and worst-case scenarios need to be developed, and the company needs to be absolutely certain it can afford the worst case. In the worst case, all the funds are lost and all the company gets is the ability to write off the loss. The amount can be predicted as long as the company has created a withdrawal plan that includes triggers indicating when to end the effort. Without this plan, the loss continues, often until company-wide failure. Therefore, the amount needs to be calculated, discussed at the highest level, and agreed upon during the planning stage – not after.

If a firm cannot afford the worst-case scenario, the firm should not open the branch. If the firm still opens the branch, doing so is not like investing in stock but more like betting on a horse race. A company that can afford the additional losses may elect to continue. If the company cannot, the company should withdraw immediately. More than two modifications in acceptable losses should signal the need for a complete review of the entry strategy and likely the initiation of the withdrawal plan. Withdrawing before the financial loss threatens the home office is an imperative risk management measure.

It is critical to understand the impact of the cash drain on a firm during the several years it may take for the new office to generate positive cash flow. The contractor may consider the outflow of cash an *investment*, but the firm's credit grantors (bank and bonding company) may consider the outflow a *loss* until proven otherwise. Failure to properly estimate and widely announce anticipated ongoing cash needs may lower a company's credit rating. Cash flow struggles will also delay payment of accounts payable and, if it gets bad enough, put a contractor out of business. The bank and bonding company should be advised in advance of expansion plans and hopefully approve of the investment. They need to understand and acknowledge any potential or planned losses in the initial years.

6.7   The Wisdom of Withdrawal Planning

Research has shown that expansion into new geographic work areas is the second most common cause of construction business failure, so it should come as no surprise that withdrawal is a strategy seldom considered in construction company planning sessions. Yet, withdrawal from any strategic business move is a valid, businesslike, and often necessary component of an overall business plan. The reason it is seldom considered is that so many construction businesses are an extension of the owner, and there appears to be some stigma or ego impact associated with retreat. There is no room in the business world for this kind of thinking, particularly in the high-risk construction industry. To manage risks, an organization plans, and the planning needs to include the formulation of viable alternatives. Every strategic decision should include withdrawal as an important alternative. Although a contractor might worry about how withdrawing will make the company look in the marketplace or to peers, the perception really doesn't matter. It would make no sense for the plan to state, "We will stay here at any cost in order to look good to our peers." Can any business afford that attitude?

Withdrawal from a market entry should be orderly and businesslike, with all loose ends tied up, in order to leave the area with as good as possible relations with other businesses, groups, and individuals. Maintaining good relations in the area in question is important in the event the company has a project in the area in the future or the company decides to attempt reentering the area at some point. In a withdrawal situation, the truth can be a valuable ally. If the competition was too tough for a regional office to succeed, why not be a *"class act"* and compliment peers on the way out? For example, the firm could create a press release stating something like the following: *"We find that the fine contractors in this area are well equipped to serve the construction needs of their clients. Our participation in this market, while welcomed, was not needed to the extent that would justify our continued presence.*" Whatever the reason for withdrawing, it can usually be stated in a way that helps to neutralize rumors. At the same time, frankness may gain the organization the respect it deserves for a smart business decision.

6.8 Summary

Completing a single project outside a contractor's normal area has risks because distance changes the degree of difficulty and therefore increases costs. Opening a branch has additional risks because it always takes more effort and money than anticipated, adversely affecting home-office profits. Accurately estimating the costs in a new location is extremely challenging because of so many unknowns: unfamiliar subcontractors and inspectors, labor conditions, geotechnical issues, and on and on. Some of these things can be researched, but until you have actual experience with them, you cannot know everything about them.

Gaining this necessary experience in a new territory enables the contractor to effectively bid and obtain profitable work and can be accomplished by obtaining small projects, *testing the waters*, and learning about the risks to be aware of. Monitoring the performance of these smaller projects much more closely than the company would for similar-sized projects in their normal working area provides the organization with an opportunity to experience the risks and learn about conducting business in the new geographic area in a way that minimizes financial risk and the exposure to the overall company.

Be absolutely certain about the ability and financial strength to expand before establishing a branch office. Do not underestimate the magnitude of the planning effort and cash outlay. Establish milestone indicators that, if not met, trigger a complete and objective review of the expansion plan, and always be prepared to implement your withdrawal plan.

7

Changes in Type of Construction

By now you should realize that our intention is to reinforce the reality that contractors become successful by doing a certain type of construction in a certain size and in a certain area. Our research indicates occasional projects that lose money are often different from the norm in terms of construction type, size, or location. Success in doing construction work of one type, size, and in a specific area is not an indicator that an organization will be successful in any other type, size, or location of construction work. We are not suggesting that a contractor must resist the idea of expanding into other types of construction, but we are emphasizing that expansion comes with serious business risks. These risks are great enough to have caused major problems for many previously successful companies.

It is critical to understand that changing the type of work a construction enterprise performs regularly and profitably is a significant business event that always includes serious risks. A construction firm seldom changes its entire business focus from one major type of construction specialty to another. Contractors rarely switch from building roads to building houses. However, expanding into additional types of construction work they have no experience with is unfortunately quite common.

7.1 Reasons for Changes in Type of Work

Each construction organization has decided on or been drawn to a certain mix of projects, which we refer to as a company's service *mix*. These are types of work the company can consistently produce at a profit. Deciding to engage in a different type of work than the type of work a firm normally does is a major strategic choice that involves a change in the relationship between the company and its environment. Many construction professionals think that a construction firm succeeds by matching the type of work it will do with the needs in the marketplace. While this is logical and, from a marketing point of view, correct, it is impractical in practice. A construction organization profits by matching its experience and capabilities with the projects it will pursue. Construction enterprises do not have the luxury of going after the type of projects they *want*. They have a financial necessity to go after the type of work they have proven experience and capability to make a profit with. Every construction enterprise has specialized expertise and capabilities they have profited from. This is and should be

The Business of Construction Contracting: Schleifer's Guide to Financial Success, First Edition. Thomas C. Schleifer and Aaron B. Cohen.
© 2025 John Wiley & Sons, Inc. Published 2025 by John Wiley & Sons, Inc.

restrictive, in that engaging in a type of work that does not specifically match a firm's experience and capabilities seriously reduces the probability of profit.

Diversification is a typical reason for adding one or more types of work. The motivation to diversify into other types of work is linked to the objectives of the firm. For example, to increase sales, to increase the value of assets, or to maintain staffing levels. Often, the driving force behind diversification is the desire for growth. Growth may result from positive strategic decisions or from defensive decisions. Whatever the reason, all diversification involves risk and uncertainty.

The ability to grow or even just sustain an organization depends on the available work in the marketplace and the amount of work an organization can perform – its capacity. Another typical reason for diversifying the types of work a firm completes is a shortage of the firm's primary type of work. A close second reason is to hasten the growth of the company. Sometimes, the decision to diversify begins with an opportunity, such as a good client or friend having a job to give out that is not exactly in the contractor's line of work but is perceived as close enough.

Whatever the reason for a change in the type of work, the change needs to be approached as an entirely new field of endeavor because that is, in fact, what it is. If the organization has no, or limited, experience performing this type of work at a profit, the risks of pursuing a new kind of work are the same as those present when starting a new business. However, contractors commonly say, *"We have been in business for years. We know what we are doing. It is a little different type of work, but it is in our field."* The problem with this attitude is that the risks are not recognized or minimized, and therefore present an even greater danger. With this attitude, the contractor may not even be cautious, may not make appropriate plans, proceed with appropriate caution, or engage in appropriate monitoring. When this happens, the contractor will not likely take needed action until after a problem develops, exposing the company to unanticipated and unnecessary risks.

7.2 Lack of Experience

The exposure to risk comes from a lack of experience (Figure 7.1), which we have discussed in prior chapters in terms of changes in project size and geographic area. When deciding whether to take on a new type of work, the question should not be, "Can you produce the work?" Rather, the question needs to be, "Can you produce the work at a profit?" The lack of experience regarding new types of work may be more significant than the lack of experience regarding changes in project size and geographic area because taking on a different type of construction can be like operating in a whole new world. Many firms will attempt to perform the new type of work in the same way they do their normal type of work – with the same equipment, the same labor force, and without additional training. Competitive success is achieved through people, and proficiency and specialized skills are critical. Consequently, one of the most obvious implications of changing the basis of competitive success (by changing the type of work) is the importance of having a workforce with adequate skills to get the job done correctly and profitably.

In the construction industry, the possible changes in the type of work are too numerous to list. As one example, a road builder might decide to bid on a commercial parking lot, and many road builders did just that after highway work slowed down or stopped in their area. Often, the results were disastrous. To make such a switch, a prudent contractor should expect a significant learning curve and

Figure 7.1 Contractors tend to be more specialized than they realize, and prudent contractors stick to the type of work they know best. *Source:* ArpPSlqee/Adobe Stock Photos.

should try a small project first. Only after examining the results and ensuring that the outcomes are positive should the contractor take on another commercial parking lot project. There is no way to accurately estimate the time or cost required to gain the necessary experience, but too many companies have made a move like this one without even knowing whether they can afford the learning curve. Compounding this risk is the fact that a firm cannot get the experience required to do the work without first pricing and bidding on the project. Obviously, with no previous experience in the new type of work, it is very difficult, if not impossible, to accurately price the work. In other words, you cannot price the work accurately without doing the work, and you cannot do the work if you cannot price it accurately. This is the essence of the exposure in attempting a type of work the company has not done before.

Some contractors have tried to overcome the estimating problem by entering into a joint venture with an experienced firm (Joint ventures also have risks, which are discussed later in this book). Being in a joint venture does not necessarily teach an organization all it needs to know in order to successfully price and perform the new type of work. Additionally, if the experienced partner is pricing the work, the inexperienced partner has no way to confirm the accuracy of that pricing. Along these lines, if each partner does its share of the work independently, both may learn little or nothing about each other's field or from each other's experience. More than one contractor has successfully completed a joint venture project of a type of work the contractor was unfamiliar with and was unable to make a profit when the firm took on a similar project independently.

To return to our earlier example, the road builder believes it is more than capable of estimating the excavation, site work, and concrete work required for a commercial parking lot. Nevertheless, the organization has most likely never actually built the much smaller and different type of curbs, pavement, and walkways. Furthermore, the organization would not have the necessary experience in the mechanical systems for entrance and exit facilities and therefore may struggle to accurately price and build them. Similarly, the firm would understand little about working in a populated area instead of the open right-of-way for road building they are used to. The project may include traffic control, rerouting public foot traffic, and providing alternative parking.

Even if the organization overcomes the lack of experience at the bid stage, the organization will still face the challenge of constructing something it has never built before. A contractor's success depends on experience. (*Luck helps.*) Even experienced people find it challenging to coordinate the shop drawings and manage the interfaces and physical space problems for all of the systems in a complicated project, for instance, like a treatment plant. In numerous instances, contractors with no treatment plant experience have been unable to complete their early attempts at treatment plant work. It is the same with any type of project that an organization has no experience with. In some cases, the contractors were able to complete the work but did so at a loss. Even experienced contractors must excel at serious planning and organized preparation in order to complete complicated projects at a profit. A contractor has almost no chance of making a profit on a complicated type of work when the contractor has no experience successfully pursuing, capturing, and producing that type of work. Yet contractors attempt it all the time.

7.3 Subtle Differences

Some differences in types of construction are subtle and difficult to recognize without having performed the types of work before. For example, many building contractors think of all building projects as pretty much the same until they take on their first hospital job. The lessons learned when a firm completes the first project in a new type of work are often severe. The firm finds out that hospital projects seem to refuse to get done on time, are impossible to schedule, and require putting three feet of mechanical systems in a two-foot space above the ceiling. The differences between a hospital and a residential high-rise are similar to the differences between an icebox and a refrigerator. *The icebox and refrigerator seem similar on the outside and both keep food cool, but one is far more complex than the other.*

The increased exposure to risk is not exclusively associated with more complicated types of construction. For example, significant experience with very complicated types of construction does not mean an organization is automatically equipped to produce less-complicated work at a profit. The issue is not the *degree of difficulty* but the *degree of experience* in the type of construction. The various types of construction are different, and experience with one type of construction does not imply ability in another, even in very similar types of construction. *A world-renowned baseball pitcher may not be very good at softball pitching.*

7.4 Recognition of Your Specialty

We are not suggesting that a contractor needs to be smarter or more skilled to do one type of construction work than to do another type. The reality is much simpler. Every successful contractor is a specialist. The type of work an organization has completed successfully and profitably is the organization's specialty. Often, an organization has more than one specialty, and each should be identified and recognized as distinct. Otherwise, a firm might take on another type of work without realizing the firm lacks experience in that work and will then wonder why the project was so difficult or resulted in a loss. This situation is particularly lethal when the project happens to be larger than the projects the firm typically completes.

A construction organization grows and becomes successful by perfecting the skills needed in the organization's specialty. Organizations learn the nuances of their specialty to the point that they can do the work better and bid it better than less-experienced competitors. Too often, construction professionals fail to realize how specialized their fields are. An organization may believe that it understands the entire industry and how it works, but when contracting to complete a project, the organization is only safe within its area(s) of experience.

It is not difficult for a construction organization to determine and understand its specialty. However, many firms do not, because they do not understand how specialized they actually are. Too many say, *"Construction is construction, and we can build anything."* This thinking is particularly prevalent among those who do not fully understand where their expertise lies. Many construction organizations are not aware of or have not identified which areas of their operational experience are strong, which are marginal, and which are weak. Without a clear understanding of their expertise, these firms do not know or measure which types of work are contributing the most to the firms' success. Almost imperceptible variations in the type of work undertaken can make a big difference in a contractor's ability to profit from a project. Just ask a hard rock tunneling contractor about the results of its first attempt at soft earth and mud tunneling.

In this section, we have discussed small changes in the type of work performed because it is easy to believe that small differences do not present a risk or do not present the same degree of risk that big changes do. The reason is that small differences are so much more difficult to recognize than bigger, more obvious differences. Another exposure is that some people recognize small differences but believe that they do not matter. We have seen experienced plumbing contractors take on Heating, ventilation, and air conditioning (HVAC) contracts, seen electrical contractors get building contracts, and seen road builders take on dredging contracts. We have experienced a long list of contractors that took on work completely out of their specialties and lost money, but we have also seen an even longer list of contractors that took on work in their general fields but outside their specialties and lost money. Some returned to their specialties, while other firms did not survive the experience of branching out. Hopefully, these examples will convince even skeptical readers that experience and expertise in one type of construction does not imply that the experience and expertise will apply to even a slightly different type of construction.

Case Study: Forced Out of Business

A sewer contractor had been in business for 12 years and did about $20 million a year profitably before it had its first losing year. That year, the contractor lost money on its two largest jobs, which were both within the firm's normal size range and normal geographic work area. When the losses on these two jobs caused cash flow difficulties, the contractor was late paying subcontractors and suppliers, who eventually complained to the project owner. The project owner, in turn, notified the contractor's surety. A subsequent investigation revealed that the jobs appeared to be straightforward underground sewage collection projects, and the contractor pointed to a string of similar jobs that he had made money on. A closer review of the previous projects revealed something that surprised even the contractor. All of the successful projects – and all of the large projects had been precast concrete pipe gravity sewer jobs. Several projects included a small amount of ductile iron pipe force mains, and none of these jobs had been nearly as profitable as the exclusively gravity sewer projects. The firm had completed two small, exclusively force-main projects two years prior, and both projects lost money, but because they were minor projects, the company was profitable for the year.

The projects this company had bid on during its first 10 years of business were exclusively gravity sewage systems. In the last two years, the firm had taken on work that consisted of gravity sewer systems, force-main systems, and a combination of the two. The contractor was busy building a successful construction business and had not realized that it was not successful at producing force-main projects. Because the firm was new to force-main projects, it had limited experience in bidding for that type of work. It was only when the firm's financial difficulties came to light that the contractor realized the two biggest projects that were losing money were both force-main jobs, and that his bids were considerably below the second bidder. Several other losing projects were combination gravity and force-main systems.

The contractor and his employees referred to the company as a sewer contracting firm because the firm had built only sewage collection systems. However, this company was, in fact, a *gravity-sewer contracting firm* because that was the only type of construction work the company had ever performed at a profit. (While there is not necessarily a category referred to as "gravity sewer contractors," that is all this firm had attempted in its first 10 years in business.) Unfortunately, the firm did not recognize its specialty until it was too late (Figure 7.2).

Our investigation revealed other reasons why the force-main projects were unprofitable. Every prior gravity sewer project had passed its final alignment inspections and leak tests, whereas none of the force-main projects passed any inspections the first time. Most of the force-main ductile iron pipe had to be re-excavated, realigned, retested, and backfilled again. It was also necessary to repair the majority of rubber gasket joint connections. The crews, who thought they knew how to place ductile iron pipe, were mistaken. They and the managers did not know, or did not care, that the specifications required the ductile iron pipe to be placed from manhole to manhole in a perfectly straight line. It was not. The pipe actually wandered back and forth from one side of the trench to the other and, in some cases, moved up and down. Further, the majority of the rubber gaskets at the pipe joints were improperly installed and leaked under pressure tests.

These circumstances are difficult to believe, considering the organization had been successful for more than a decade. The only change was the type of pipe, and the contractor did not consider that change to be notable until it was too late. The losses were larger than the company could survive, even though all of the exclusively gravity sewer projects were profitable. The company had little choice but to go out of business. One of the authors of this book worked personally with this company and can assure readers that the firm's owner was smart, honest, and sincere and had done an excellent job building a successful construction company. The contractor said he had every intention of including force-main work in his operations, and he was shocked at how difficult it was to accomplish this aim. He added that he now realized that the firm's unprofitable first attempts to complete a force-main project should have been a

Figure 7.2 Installing force-main sewers and gravity flow sewers may seem like similar work, but an experienced contractor knows that these are two very different types of construction. *Source:* fefufoto/Adobe Stock Photos.

warning, but the firm was so busy growing that he really did not take notice. Then, when the firm took on two big force-main projects at the same unit prices as the similar prior losing force-main projects and tried to perform the work with the same crews, they were unsuccessful. The reason for the company's downfall was clear to the contractor now.

Case Study: A Mechanical Failure

The largest mechanical contracting firm in a major metropolitan area had been in business for more than 50 years and had three divisions generating above-average profits. The HVAC division operated its own sheet metal shop, and the mechanical division included a fire sprinkler operation with its own pipe fabrication shop. The service division provided contract maintenance services to companies throughout half of the state and handled all the projects that were not large enough for the other two divisions. The service division also enabled considerable efficiency and increased profits in the other two divisions because it did all of the start-ups, punch-list work, and call-backs for the other divisions. This arrangement made sense because the service division had over 100 fully equipped service trucks, which were operated by highly skilled tradespeople.

Though the HVAC and mechanical divisions were both larger than the service division, the service division made a much higher profit percentage. The service division was growing rapidly, signing more and larger multiyear service contracts, and taking on bigger and bigger jobs that the other divisions had to turn away. Nevertheless, the service division did not receive much of the top managers' attention, because the other two divisions brought in the lion's share of total sales.

Eventually, the top managers/company owners retired, and the next generation of the family took over. The new leaders felt they were overworked because of all the responsibilities they had over the three divisions. The leaders decided that since the HVAC and mechanical divisions represented 80% of the company's sales, the service division should be sold. The new leaders also determined that the sheet metal and pipe shops were producing products that could be bought at close to the same cost; therefore, the sheet metal and sprinkler pipe shops could also be sold. Contributing to the decision to sell them was the fact that the shops, as well as the company offices, were located on expensive real estate that could be sold for a huge profit.

After the service division, shops, and land were sold, the streamlined organization located its estimators and contract managers in an office park. The managers were pleased with the huge reduction in overhead costs and were able to increase the sales of the now-merged HVAC and mechanical divisions to replace the sales that were lost from the service division. The following year, the company was able to increase total sales by 20%.

Unfortunately, the company never had another profitable year after its reorganization. The new top managers had not been around during the decades when the company grew from a medium-sized firm to the largest in its market. But these managers understood marketing very well and had inherited good pricing methods, so they had no difficulty getting work. The managers had reasonable administrative skills and therefore were able to account for the work. What they were unable to do was to produce the work at a profit. The problem was that what they considered streamlining the organization's production had disassembled the interacting and overlapping divisions the founders had developed, which maximized efficiencies and profit. The interconnections that had been developed over decades to deliver their type of work and generate exceptional profit were lost.

For example, the new managers did not realize that because the service company had absorbed the start-up, punch list, and call-back costs for all the projects, the service division significantly improved the profitability of the HVAC and mechanical divisions. The service division was able to perform this work for the other divisions at much lower costs because it used highly skilled, permanent employees.

In addition, the new managers did not understand that the sheet metal shop and the pipe shop that were sold provided the HVAC and mechanical divisions with exceptional service that seriously magnified the divisions' productivity and profits. The internal work was always produced on time and, in some cases, overnight. Perhaps more importantly, unexpected changes during construction were handled immediately, and problems were solved quickly to avoid project delays. The HVAC and mechanical divisions operated in a very price-competitive market and captured a large share of their work because they were always able to control the work and produce it on schedule, all while maximizing profits by being their own supplier.

After the service division, pipe and sheet metal shops were sold, the HVAC and mechanical divisions could no longer produce projects at their formerly competitive prices. Ongoing projects that were profitable before the streamlining immediately began to underperform. It became apparent that even the prior managers did not understand how much the overall organization had been profiting from the efficiencies and synergies that resulted from the cooperation between the three divisions and the two shops. The root of the problem was that the top managers did not understand the organization's efficiency and true expertise. In effect, they did not understand the "type" of work the organization excelled at (Figure 7.3).

Figure 7.3   It's critical for management to understand the type of work an organization excels at. *Source:* Hryhor Denys/Adobe Stock Photos.

Three years after these changes in the service mix and type of work the organization performed, it found itself in serious financial difficulty. The company got into trouble doing what the new managers thought was the same type of work as the company had been doing for many years. However, they changed what was left of the organization to the extent that the type of work they now performed was nothing like what they did before, nor was it conducted in the same manner. The work was being performed in a different fashion through altogether different processes and methods. What they thought would be a more streamlined manner of operating was not profitable. While seldom recognized, a company's construction processes are part of its specialty that enables it to deliver its type of work at a profit. These processes can be very sensitive to change.

The risk in changing the type of work is compounded if the size of projects changes as well. The risk can be further compounded when the new type of work involves a new type of owner, such as if a company shifts from paving for private developments to paving public highways. Few understand or appreciate that the "type" of work they do is specific and identifiable, and that change creates a chain reaction that is difficult to anticipate. There is risk in going into business, risk in being in business, and risk associated with changes to a business.

7.5   Union Versus Merit Shop

A number of factors make what appears to be the same type of work quite different in reality. These factors can create problems and losses if the contractor lacks the experience to deal with them. Consider, for example, the differences between operating with a union labor force and operating with a merit shop labor force. Attempting to switch from one to the other will likely lead to a culture shock and can create serious complications and changes in costs.

If the complexities are anticipated and dealt with in advance, most contractors can survive a permanent switch, but the switch can still cost a great deal of money. A temporary or occasional switch is almost reckless. When merit shop contractors take on their first union job, they usually choke on what they consider to be restrictive work rules. This situation is particularly true for smaller projects. Therefore, the switch is not recommended at any price for small firms. Conversely, when union contractors do their first open-shop project, they may have difficulty finding the skilled workers that the contractors had grown accustomed to when working with trade unions.

Suffice to say that operating as a union shop is different from operating as a merit shop and the reasons for switching in either direction can be varied and sometimes personal. If the change is necessary, there are legal and practical issues that need to be carefully measured before attempting the switch. Perhaps the greatest risk is that the organization's managers may become so tied up in learning aspects of the new processes that they are distracted from running the business overall. Successful construction organizations become profitable by refining their methods of producing the work, and any disruptions to these routines are prone to create negative results.

We are not implying that companies should not switch from a union shop to a merit shop or the other way around. Rather, we are emphasizing that there are risks in making the switch. Therefore, before attempting to make the change, a company needs to acquire knowledge about the differences, make well-thought-out plans, and be certain that the company can afford any resulting costs. Seeking advice from contractors who have successfully made the switch is an effective way to obtain information because there is no reliable how-to manual. The most significant issue is that there will be costs, and these costs will be difficult, almost impossible, to estimate in advance. Unsuccessful switches have resulted in business failures. Therefore, caution is key.

7.6 The Importance of Knowing the Risks

Some may think that the risks involved in changing the type of construction a company does are perfectly obvious. However, the extent and seriousness of these risks involved are complex, difficult to confirm, and certainly not obvious. Consequently, too many good contractors have made such changes with disastrous results. History tells us that the risks accompanying drastic changes are often overlooked, so it is understandable that the risks associated with more subtle changes are seldom considered. Subtle changes in the type of construction attempted are nearly as dangerous as drastic changes but are often undertaken with such confidence that construction professionals are surprised and even shocked when the roof falls in.

In the construction business, a lot of money is turned over and a very small portion stays with the company. Therefore, it is imperative that construction professionals understand exactly where their organization's expertise lies and how it generates profit. Risk management includes knowing what type of construction the organization does profitably and which subcategories of work the organization does best. Only then is the organization ready to confidently move forward with any work. Then, top managers can decide how much additional risk they want to take on as they expand the company. This approach discounts the notion of "we can build anything."

7.7 Volume Versus Profit

The primary reason for undertaking a different type of work is a lack of the company's primary type of work. When not enough of the firm's normal type of work is available to satisfy the firm's appetite, managers need to be very careful about going after a type of work they have no direct experience in. Few managers recognize that when there is a lull in the marketplace, they can allow the company's sales volume to decrease. This option is always preferable to embracing the risk of attempting work outside the organization's direct experience. It is impossible to accurately measure the risk, in part because it is extremely difficult to recognize the risks involved in the unknown, let alone estimate the costs involved. The alternative to taking on a new type of work includes the risk of catastrophic loss. Some people would say that taking work outside of the firm's experience is a defensive reaction. Because of the number of losses and failures that have resulted from a change in the type of work, this option is better described as a disastrous reaction.

A market slowdown can be seen as an opportunity. The organization has more time to concentrate on the existing work, and increased concentration usually increases profits without any increase in risk. The more an organization produces its specialty type of work, the better the organization gets at it. When the number of available projects levels off or decreases, everyone in the company has more time to devote to the work at hand and profits consequently grow. When managers are not rushed or overworked in an attempt to expand the company, they have time to make improvements in the work they perform, including estimating, bidding, and managing the work, which improves the chances of capturing the specialty type of work when it becomes available. Having time to fine-tune skills used in an organization's specialty can increase the potential for profit, making the sacrifice of decreased sales not only a risk-free alternative, but also the most prudent alternative.

Most people will agree that earning a greater percentage on a lower volume of work while experiencing little or less risk is real growth. If the market does not rebound quickly, delaying the decision to attempt a different type of work is still beneficial. The reason is that when a company experiences market-forced reduced sales, downsizing tends to improve cash flow, in which case the company will at least have greater financial reserves. If the company eventually decides to take on the risks of expanding into a new type of work, the company will do so with increased cash reserves.

7.8 Withdrawal Plan

As discussed previously, a primary element of risk management is creating a withdrawal plan. This plan needs to be created before an organization attempts a new type of work. The organization should create a withdrawal plan, which some refer to as a supplement to the business plan, that details how the work will be captured and performed and what will be done if the effort is unsuccessful. The process of crafting a withdrawal plan starts with determining how long to continue attempting the new type of work if it is not successful. Managers must also determine the milestones they will use to measure the success of the effort. There is no need to be too optimistic. Making a profit during the first or second attempt at a new type of work may not be possible, and profit is not the only measure of success. The company may be trying to penetrate a market or be on the ground floor of a developing or expanding type of work. It is reasonable to expect to lose money on the first job or two. Managers need to determine how much of a loss the company is willing to absorb and can afford.

Managers should also evaluate how a loss will affect the organization's financial statement and credit. Obviously, cash must be available to pay for expenses. A failure to accurately estimate the potential costs and to determine how to cover them will decrease a company's credit standing and could even put it out of business. These considerations should be included in the withdrawal plan, which should drive every strategic decision. Withdrawing from performing a type of work can usually be accomplished with few people, if any, outside the company taking notice.

7.9 Summary

If managers are convinced that their company must move into a new type of work to maintain volume or enable growth, they should first evaluate all of the risks and then proceed with caution. We recommend testing the water with a smaller project to minimize exposure to risk. If the smaller project is not successful, the organization can withdraw gracefully (in accordance with its withdrawal plan) and plan to try again or plan to try something else. It is never appropriate to bet the whole company by starting with a large project in a new type of work.

8

Changes in Key Personnel

Most of the built environment in the United States is put in place by small and midsize construction enterprises, and the majority of construction companies are closely held businesses managed by some or all of the owners. In the majority of smaller construction companies, the key people are usually easy to identify. They are usually the owner(s). In midsize and larger companies, the key people are the owners and top managers. The term key people can be defined as the individuals who are critical to the success of the enterprise (Figure 8.1) and, as such, are not easily replaced. Senior managers may or may not own a small portion of the business. Midsize and large enterprises may have as few as three key people, with each in charge of one of the three primary functional areas of the business. In smaller businesses, one person may be in charge of all three functional areas.

As discussed in prior chapters, the three primary functional areas of a construction enterprise are getting the work, doing the work, and accounting for the work. In each successful construction organization, someone has direct responsibility for these functions. Any weaknesses or underperformance in any of these areas in a construction organization can be traced back to the person in charge of the relevant functional area. A weakness in any of these areas can cause a construction business to fail. The loss of the key person who is responsible for just one of these areas can and has caused many companies to flounder.

This chapter focuses on the loss of one or more of the key people without whom a company cannot function properly. The information in this chapter may apply to middle managers, who are a very important part of an organization, but, in the majority of cases, changes in senior managers are almost always critical.

8.1 Breakup of a Partnership

We have observed that many people who jointly founded and successfully ran a construction company and then later split up were not nearly as individually successful after parting ways. Very often, it takes the contributions of more than one person at the top for a construction business to flourish because many people are skilled in a limited number of areas. If one partner is excellent at construction operations (i.e. doing the work) but not the other areas, the business may fail without the former partner who is skilled at marketing and sales (i.e. getting the work) and managing the business (i.e. accounting for

The Business of Construction Contracting: Schleifer's Guide to Financial Success, First Edition. Thomas C. Schleifer and Aaron B. Cohen.
© 2025 John Wiley & Sons, Inc. Published 2025 by John Wiley & Sons, Inc.

Figure 8.1 Every contractor has key people who are the primary reason for the company's success.
Source: tuiphotoengineer/Adobe Stock Photos.

the work). The opposite is equally true. A person who can get the work may not be skilled at doing the work and, therefore, needs a partner who has that skill set. A construction owner can of course hire someone to do any of these jobs, but typical entrepreneurs tend to be doers rather than managers, and most owners have not had the formal management training needed to manage staff effectively. It is difficult, if not impossible, to find someone to do a task as diligently and conscientiously as an owner would do the task. Further compounding the issue, most employees who have what it takes to lead a business already have plans to do so and are difficult to retain long term.

Another issue regarding the breakup of a partnership is that neither individual is likely to have experience single-handedly managing the original size of the business. Thus, to attempt to operate at the original scale is to assume that the missing partner did nothing or that the remaining partner can do, in his or her spare time, whatever the other partner did in their full-time capacity. Logic suggests that the ideal approach to manage risk is to reduce the size of the business to no greater than half the original size. However, after a partnership breakup, few contractors will accept such a reduction in volume under any circumstances, and cutting a business back by half may cause more problems than it cures. A prudent alternative is to limit growth for a couple of years to determine whether success continues in the absence of one of the key people. Assuming that nothing will change defies reason.

8.2 Founders and Succession

Clearly, the loss of a business founder is a difficult blow to an organization. There are no commonly accepted succession models for closely held construction companies. And few develop succession plans in advance. Most construction organizations are shaped by owners as the business progresses and grows. Because many sizable construction companies were started by relatively young men, it is not unusual to find 30- to 40-year-old companies still being managed by the founders. These owners often struggled early on to discover exactly which management processes, procedures, and styles to apply. Over time, these owners learned which approach worked for them.

The founder's values and business methodology are a big part of the reason that companies are able to grow and prosper. After many years and the addition of numerous middle managers, the founder's values may become all but hidden. However, they can usually be found embedded in the operating strategies of the business as long as the founder continues to participate. Unfortunately, the founder's values are often seen by future leaders as roadblocks to the company's further growth and are often described as *old-fashioned*. Experience tells us that when successors take over, changing the values too much can result in the company evolving away from its formula for success.

We are not implying that companies cannot succeed after the loss of the founder. However, the reality is that many construction enterprises do not survive succession. We believe that most of the businesses that failed after founder succession could have survived if they had developed proactive plans for transition and had a healthy, objective understanding of what made the company successful. Many successors immediately change anything and everything in the name of modernizing the organization, but doing so can cause the organization to underperform.

8.3 Inactive Founders

Another issue is when a founder becomes inactive in an organization after succession. When the founder is in the background, the new leaders are often reluctant to change much if anything, which usually allows the business to continue smoothly until the founder dies. A strong founder can influence a company long after becoming inactive. Another phenomenon observed is that the loyalty of long-time employees often prevents them from leaving the company as long as the founder is alive. The influence of long-term employees helps keep the founder's values alive under new management. In some cases, the new management lacks the strength or fortitude to change things too drastically while the founder remains in the background. However, when changes occur, it often results in the departure of long-term employees.

We have also seen cases in which companies experienced successful management succession but were negatively affected by actions of the founder's estate after the founder died. The succession of any top manager, particularly a company founder, is a critical juncture for any closely held construction enterprise. At this point, failure is preventable but requires a lot of planning and effort. Unfortunately, planning and effort don't always occur in this industry, which largely does not see succession as the high-risk event that it is.

When the founder or a previously active partner is no longer around, a company is more susceptible to changes. We have observed that large, established businesses approach a leadership transition with no more planning than a retirement dinner. Some top managers in construction companies believe that a successful enterprise will continue to be successful by default. In reality, if any one of a firm's key people leaves the company, it is in jeopardy until a replacement is found and that replacement is tested. And testing takes a minimum of a year. Even a replacement with equal skills and talents will lack the missing key person's connections and the loyalty of followers. These intangibles are lost to the company and are part of the reason that key people are so extremely difficult to replace.

If the replacement does not work out, the replacement process must start over. All the while, the firm's bank, bonding company, clients, and employees worry about the company's sustainability. The risk that comes with the succession of key personnel cannot be overstated.

Case Study: Succession

As we present the following case study, keep the three functional areas of a successful construction enterprise in mind. It is essential to understand the importance of these functional areas in a construction organization. Additionally, it is essential to understand that a change in the person responsible for any of the three functional areas puts the company at risk. Therefore, all changes should be treated with great caution.

Now for the case study: Two partners started an industrial construction company. Not long after they began, they decided to bring in a third partner to run the accounting department. The new partner was not given equal ownership right away, but the original partners wanted someone with an ownership interest in this critical area because they had experienced some close calls in the past as a result of inaccurate accounting during a difficult growth period. Over time, the company grew to $200 million a year in sales. The company was financially solid and working in a growth market. The new Chief Financial Officer (CFO), who was now a one-third owner of the company, was a good leader with a conservative approach to the business.

After being in business for 20 years, the company had been offered the opportunity to buy a firm that provided the design and engineering for some of their auto industry clients on projects the company produced. The design firm was not very profitable, so the partners thought they could buy it at a reasonable price and increase their competitive edge by expediting the designs for their work. The two original partners liked the idea. The third partner, the CFO, was passionately against the deal because the only option was to pay in cash, and the construction company still carried some debt from having grown rapidly for so many years. The CFO said that using all available cash would leave them at the mercy of their creditors for a number of years. Even though they were in a good market, the increased payments on their debt would consume a substantial portion of their annual profits.

The CFO refused to agree to the risk. The pressure to purchase continued, and the issue finally had to be settled by the board of directors. The board decided against the purchase largely because of the CFO's financial presentation. The CFO was well respected by all concerned and had an excellent understanding of the business, particularly cash flow. He remembered both the good and the bad years and understood that, in a cyclical, high-risk industry such as construction, a modest cash reserve and some unused credit are not a luxury but a necessity. He kept watch over every financial aspect of the business, and during his presentation, he explained to the other partners and the board that though the business was financially sound and in a growth market, increased competition was eroding profit margins.

The company's primary type of work was constructing and renovating industrial facilities for the auto and steel industries. Though plenty of that type of work was available, road building and commercial construction had slowed in the area. Consequently, other types of construction firms were becoming desperate for work and were bidding on industrial projects.

Increasing competition was not the primary reason the board found the purchase too risky. What tipped the balance was the CFO's graphs showing how the company's cash flow decreased eight years prior because of a market decline following a dip in the national economy and a lengthy steel strike. The company had remained marginally profitable during that period, but their cash flow had become dangerously negative because of suspended projects for which retainage could not be collected timely and because of slow-paying clients. Numerous canceled projects had added to the profit problems. What few members of the company remembered was that the CFO had saved the day by insisting that overhead be cut immediately. His charts showed that, without their reserves of cash and credit at the time, the company could have gone under.

Two years after the board rejected the purchase of the design firm, the CFO had a serious illness that forced him into retirement. The company hired an accountant from its auditing firm because the accountant was very familiar with the company. Everyone was happy with the selection, and because the marketplace was still strong, the future looked excellent. Six months later, the partner in charge of construction again proposed purchasing the design firm, which was still available. This partner claimed that the company needed to make the purchase now more than ever to improve profit margins.

The other partner was skeptical, and the decision again went to the board of directors. The partner in charge of construction presented a good argument for making the purchase, claiming it would lead to a minimum 2% increase in field productivity because of better design service and faster delivery of design modifications. The new CFO reported that he had studied the figures from the field people, as well as the purchase price information from the design firm. He claimed that they could definitely afford the purchase. When asked if the purchase would tie up all the company's cash and borrowing power, he reported that it would be easy to increase the firm's line of credit. No one asked how the CFO had made his calculations.

As it turns out, the new CFO had taken the company's profit over the previous two years and had calculated an increase of 2% across the company, not just in terms of the work directly associated with the design firm. Further, no one questioned the fact that the design firm was involved in only auto industry work, which represented one-third of the work that the company performed at that time. With this oversight and the failure to ask how the CFO had made his calculations, the board approved the purchase on the condition that a certain increase in the line of credit could be negotiated before the purchase.

After the purchase, the company discovered that the design firm was even less profitable than originally thought and some additional cash had to be put into it. On the other hand, purchasing the design firm increased the company's efficiency and increased profit margins on the work involved even more than originally projected, so the purchase was considered a success. Two years later, however, during a mild recession in the US economy, the auto industry slowed down. One of the company's major auto industry clients ceased all construction and renovation work. This loss in work cut the design firm's volume almost in half. Less than six months later, three major auto industry projects the company was about to start were delayed indefinitely. Steel industry clients were also slowing down and cutting back on their construction activity, and five major industrial capital improvement projects that were under construction stopped without notice. The company suddenly was losing money daily. In spite of valiant attempts to cut overhead and find work in other areas, the company could not reverse the trend. The company's limited cash flow made it impossible to pay the interest on loans, let alone the principal.

The CFO's attempts to renegotiate and restructure the company's loans failed because the bank was concerned that the company might lose their bonding capacity. The bank wanted a written commitment from the construction company's surety carrier that all necessary bonding would continue to be available to the company. The bonding company would only say that each project would be looked at independently. The design firm was eventually given back to the original owners at no cost, and the company was considering the option of filing for reorganization. A buyer was found before that option was carried out. A change in one key person had cost the partners their business because that person, the original CFO, was the only one who really understood the long-range risks.

The partners realized the cause of the failure only after the surety's consultants (the author) reported that the new CFO's loan payment calculations were misleading. In the loan payment calculation the CFO presented to the board, only the interest portion of the loan payments was reported as a cost to the company. The CFO made this claim because he said paying the principal portion of the loan was not literally an expense to the company. However, from a cash flow perspective, the distinction was meaningless because the loan agreement stated that the entire payment (interest and principal) had to be made each month. The CFO's projections would not likely have been accepted by the board had the full loan payment been included in the CFO's calculations.

8.4 Replacement of a Team Member

For all intents and purposes, changing the key personnel in a construction enterprise creates a new company. The reality is that a formerly successful construction company with a new management team has yet to prove that the company can continue operating profitably (Figure 8.2). The new management team may very well be successful, but it should not be assumed that it will be successful. Many believe that each person in an organization works independently and does not affect other parts of the organization.

Figure 8.2 The training of future leadership is critical to the continued success of the organization. *Source:* TommyStockProject/Adobe Stock Photos.

However, the reality is that in a relatively small organization, such as a closely held construction enterprise, the functional areas of responsibility are interdependent, not independent. Therefore, replacing the person responsible for one of the functional areas affects the entire company.

Just adding a person to an existing management team alters the team regardless of the size of the team. For better or worse, the team is not the same. The definition of *team*, as presented in the book *Wisdom of Teams*, sheds light on the impact of changing a member of the team: "*a team is a small number of people with complementary skills who are committed to a common purpose, performance goals, and approach for which they hold themselves accountable.*" The key phrases in this definition are *small number, complementary skills, common purpose, goals*, and *approach*. The likelihood is slim that all these aspects will remain the same when a member of the original team is replaced. As an analogy, if you substitute an ingredient in a cake mix, you still get a cake but not the one defined in the original recipe. Changing one person changes a team's potential.

As we already noted, to reduce the risk associated with changing a key team member, you need to reduce sales volume, or at the very least, not grow for a year or so. Reducing or maintaining sales volume allows time for the new management team to prove it can perform successfully. Absent any growth, if the new team does not perform as required, the company has the opportunity to adjust. If the firm grows during the first year of a new team, the risk is magnified.

8.5   Addition of a Key Person

As a company grows, in addition to replacing departing employees, it must continually add to its management team. During rapid or sustained growth, new people are regularly placed in key roles, meaning that the company is repeatedly a new, untested organization, performing a greater volume of work. A successful new team depends on whether they can function together to accomplish the mission and objectives of the company at a profit. This cannot be known until the new team functions together for some time. The authors believe that anything less than a year is an insufficient test. A company may have more objectives than to be profitable, but they are meaningless unless the company generates a profit. To succeed, a company must earn funds that exceed costs. A group that does not achieve this goal can call the organization whatever it wants, but it is not a *business*. A business is defined as operating for a profit. A not-for-profit business is a charity.

Few growing companies take the time to analyze their situation after adding key personnel because it is difficult to isolate specific variables when the entire company is occupied with growth. If problems develop some years down the road, there may be new people or circumstances to blame, but the former decision-makers are nevertheless responsible for how their decisions to expand the organization affect the company's future.

8.6   Management Dilution

A similar and potentially more serious problem that construction enterprises face during periods of rapid growth is what we refer to as a change in key personnel by *dilution*. This occurs when a significant number of employees are hired into a closely held company, and teams that have been working together for some time get *diluted* when newcomers are introduced into the teams. Contractors are often determined to expand their businesses because of the loyal team members who have worked so hard to generate the organization's current success. However, what can happen when teams are expanded is that existing team members begin to feel less appreciated or displaced. Often, a company outgrows the existing middle management team's capabilities to the point that new managers with more or different experience are required and introduced into the teams. Hiring new people and placing them as managers over existing managers can cause the existing managers to be offended and feel like their years of loyalty are being ignored. Loyal long-term employees often experience suspicion, jealousy, and even resentment. Some valued and potentially essential employees will decide to leave the company.

Employees choose to work for a company for various reasons, including the size of the organization, regular access to the boss, and the office atmosphere. In the construction industry, there is also a certain amount of what we might call *hero worship*. Charismatic entrepreneurs are able to generate loyalty in employees, motivating them to work extreme hours and put in staggering effort in order to work with the entrepreneurs.

As a company grows, new managers are placed between the entrepreneur and loyal, hardworking employees, resulting in decreased contact with the boss. Our research shows that a company that doubles its size in three years will inevitably lose formerly loyal employees. The number of employees who leave varies. Sometimes the majority of the firm's core employees leave as a result of significant growth. The negative impact is greater when the owner is unaware of or does not understand this phenomenon, and therefore does not expect it to occur and consequently takes no steps to prevent it.

Inevitably, some employees will leave as an organization grows. However, change is not always bad and some things will be gained. Just as a plant that has outgrown its container must be transplanted if it is to continue to grow, an organization that has outgrown its infrastructure, style of management, or people must also make changes. Failure to do so will lead to a variety of other problems. We call these problems *dilution of key personnel*.

Case Study: Increase in Organizational Size

This case is about a prosperous small construction organization that expanded into a midsize firm. Here, our definition of *small* is that the contractor has some level of control over all the work. His current sales were $25 million. In a midsize firm, authority has to be delegated, and the owner(s) transitions from doing the work to managing the business. As the owner explained it, the company was expanding into a 7,000-square-foot office space, which was a big increase from the less-than-2,500-foot office they were in. The company had significant opportunities they were going to take advantage of and planned to double their size in 16 months and potentially double again in four or five years. The owner asked us what we thought about his expansion plans. We described some research we were working on that indicated an expansion of that size would have a significant and potentially unexpected impacts on his organization.

The organization consisted of a tight-knit group of people the owner had worked closely with for years, and he explained that his loyal employees were precisely the reason he was able to expand. We mentioned that if he more than doubled the number of employees, he would not have the same personal interaction he currently had with his employees and that his current employees were accustomed to. Consequently, the work dynamic among them would likely change. We explained that during substantial expansion it is common to hire senior people and that some of his current employees would undoubtedly have to report to them. The contractor felt that would not be a problem because his existing employees were a loyal and close-knit group.

The contractor's first step was to hire an executive vice president to take some of the management load off of himself so he could concentrate on capturing new work. When the company moved into the new offices, the first thing the executive vice president did was reorganize the company into several departments and establish three divisions. This reorganization separated the current workforce into different groups, and only one of the existing employees was selected to head a department and none were selected to head a division. The first sign of discontent was that most of the long-term employees were complaining among themselves that they never saw the boss anymore. The boss's office was in what was being referred to as the executive suite, which the employees had limited access to.

The next time we heard from the contractor, he was upset that two of his best employees, whom he considered friends, had left the company. He called a couple of months later to say that another old-time employee had just left the company. He added that two others had left since the previous time we talked, increasing the total to five. He was shocked that it was happening just as the research had indicated.

The lesson here is that employees, particularly those at small companies, usually work at a specific firm because they are comfortable there. They like the people they work with, including the boss, and the comradery. Loyalty develops, and so does a sense of accomplishment. Employees may not be conscious of these factors. They just know they like working there. In this company, the work dynamics were totally disrupted almost overnight. Typically, the affected employees are not offended or disgruntled in the short term. They simply become less satisfied while at work, and eventually, it dawns on them that they are not as content as they were.

What few employers seem to recognize is that, in today's marketplace, their employees are often sought after and almost regularly receive job offers. These opportunities have limited appeal when an employee is content and loyal to their current employer. Job satisfaction, contentment, and loyalty go hand in hand. When the work environment changes and contentment decreases, employees are more likely to consider the ever-present external opportunities. Employee contentment is every employer's responsibility, but some employers say, "It is not my problem but theirs." In a growth scenario, there is little that can be done to prevent employee reactions, but some companies have found success in communicating growth plans to employees well in advance. The communication needs to include an explanation of why the expansion is necessary and a discussion of the potential consequences of hiring new people, senior people in particular. Existing employees also need to be assured that they will continue to be treated fairly. Communicating these issues and ideas will go a long way toward helping a company retain their valued employees.

8.7 Summary

Construction is a difficult business that requires one or more key people to oversee the three functional areas for the company to be profitable. The loss or addition of key people creates a new and untested management team, which, in many ways, can be thought of as a new company and immediately puts the business at risk. It should not be assumed that people who have succeeded separately or in other groups can succeed in a new group. Each new group has its own challenges and needs to prove its ability to make a profit. The risk is unavoidable but can be managed by not assuming that all is well that looks well. The new team needs to be proven for at least a year before the organization seeks to further expand the business and increase in size again.

9

Managerial Maturity

The exciting business of construction attracts many new start-up companies each year and has been doing so for a long time. Even the largest construction companies in the country today were at one time start-ups, and were often started by a single person. Attesting to this is the fact that most construction companies carry the name or initials of the founder. Most founders probably never expected or envisioned that their efforts would develop into the big and successful enterprises that exist today. In fact, some of today's biggest firms were started by men or women who couldn't find construction employment during slow periods or by individuals who simply wanted to be their own bosses.

Very few of these contractors started out with clearly defined long-range plans to become nationwide or multinational construction companies within a predetermined number of years. Interviews with numerous founders of large US construction companies offer insight into their modest expectations upon company start-up and their genuine surprise at the size their firms ultimately grew to. The modest initial expectations held by the founders can present managerial challenges. As the company grows in size, different managerial skills are needed to maintain success. At the various stages of growth, contractors should be cautious that their growth does not exceed their managerial capabilities.

This leads to the curious question: If large construction enterprises were unplanned, why did some start-up contractors have better management skills than others? The answer is that these organizations evolved from smaller organizations because of *business skills* rather than *building skills*. Historical information confirms that careful planning was not the driver. In fact, many of the founders of today's huge construction companies resisted growth. In most cases, it was the second or third generation of leadership that grew the firm. To be sure, most large- and medium-sized construction companies today have sophisticated planning strategies with elaborate systems and are managed by highly competent businessmen and women. But that's today. Almost all of these successful enterprises can point to a time in their history when their longest-range plan was how to make payroll the following week, and the most sophisticated system they had was the back of an envelope. Unfortunately, there were a huge number of failures for each of the start-ups that succeeded. Despite the construction industry's significance in the economy as the second-largest industry in the United States, our industry suffers from the second-highest mortality rate (the restaurant business is first) with the average start-up construction company lasting only five years.

The Business of Construction Contracting: Schleifer's Guide to Financial Success, First Edition. Thomas C. Schleifer and Aaron B. Cohen.
© 2025 John Wiley & Sons, Inc. Published 2025 by John Wiley & Sons, Inc.

9.1 Importance of Management Skills

So, why do some start-ups succeed and grow into large, long-lasting companies while so many others fail? And why does this pattern continue today? No doubt, luck has some effect. But so many failed contractors were competent constructors who worked under the same industry conditions as their more successful competitors suggest that we can eliminate luck as a primary cause. Since the best businesspeople and entrepreneurs seem to create their own luck, we need to look further for the cause of success.

The discriminating variable and real difference that we uncovered between the continuing successful construction businesses and the early and midterm failures was management skills. These are not just more sophisticated leadership skills that most entrepreneurs have in abundance. Analytical skills and thought processes are involved, as well as the patience to sacrifice short-term gains for long-term optimum results. Management skills include a certain amount of vision so that planning for and even dreaming of the future can take place. The lesson here is that because most start-up contractors are not trained in business, or management science, or not inherently gifted managers, it is critical that they develop the managerial skills to enable their companies to grow beyond the start-up stage.

Start-up contractors find out very quickly that there is a lot more to running their business than putting the work in place in the field. They need to attract and capture new work, account for the money, administer all the details associated with subcontractors and ordering materials, meet the payroll, acquire business insurance and bonding capacity, and much more. Some start-ups do not continue because the principal never anticipated all of this responsibility, or they simply did not want to do it. Some principals try to remain in the business while ignoring the details. Others address the details, but not well. This last group makes up the huge number of start-ups that don't go beyond the first five years in the contracting business. The single most important reason these companies do not succeed long-term is a lack of business and management skills (Figure 9.1).

9.2 Company Growth Phases

Construction contractors that succeed beyond the early mortality years usually go into a growth phase that can last for the life of the business. Few reach the five- or ten-year point and level off. Most construction businesses attempt to grow quite rapidly during the early stage of their development, which makes management skills the most important ingredient in a contractor's formula for success. If a business remains stable in size, it can be operated in the particular style of the founder that seems to work for them. But as a company grows, its management approach must also mature. If a business is expanding even at a modest rate of 10 or 15% a year, it will periodically grow out of its own systems and procedures, which, in turn, dictates that its management needs must change.

The senior management of rapidly growing entrepreneurial companies must simultaneously cope with endless day-to-day issues while considering and planning for the company's future strategic direction. The managers of most such companies are going through the process of building a company for the first time. This is about as easy as navigating uncharted waters in a leaky boat with an inexperienced

Figure 9.1   Management skills, rather than technical skills, are essential to the continued success of a construction business. *Source:* BalanceFormCreative/Adobe Stock Photos.

crew while surrounded by sharks. The sea is unfamiliar, the boat is clumsy, the skills needed are not readily apparent or not fully developed, and there is a constant reminder of the high cost of error in judgment.

If a construction company's maximum relatively safe growth rate is about the 15% mark, the need for management development is compounded beyond this point. When growth is rapid, management must not only manage the company's operations but also the growth process itself. Planning for and handling growth is an important factor in an expanding business, and dealing with ever-changing business conditions becomes a process that must be diligently managed. In spite of this reality, many contractors double and redouble their growth year after year with little or no attention to the need for organizational growth and improvements. Because of the stress placed on a poorly managed organization, getting more work can be as much of a problem for a contractor as not getting enough work.

Management ability is critical not only to the survival and continuance of a start-up construction enterprise, but also crucial during every growth period during a construction company's lifespan. Many struggling businesses claim good or bad luck affects their success, but luck has never been proven to be a make-it or break-it factor. Other factors that are commonly blamed for failure include working in a good or bad marketplace as well as having a good or bad labor pool. Yet, these factors do not correlate to construction company failure because there are always other construction organizations operating under the same conditions in a given area, and they don't all fail.

9.3   Limit of Managerial Effectiveness

Growth presents a wide range of challenges for any business. A construction contractor running a fast-growing company often finds it difficult to know when they have reached the limit of their managerial effectiveness. Small- and medium-sized companies often lack someone in the management team who is vocal, sophisticated, or objective enough to warn the management team of this pending problem. For other contractors, the biggest obstacle is an inability to delegate authority within their management teams. It is not uncommon to see strong entrepreneurial contractors in large companies surround themselves with weak top managers because they prefer not to give anyone real authority within the organization, and some discourage anyone from making suggestions. These same contractors then wonder why they are working so hard.

The number of contractors who have enjoyed many years of success only to fail because they let their companies get larger than they could effectively manage is staggering. This happens when a contractor's business outstrips their managerial ability or when the contractor is not able or wise enough to bring in the new management talent that the organization needs. A considerable amount of managerial and personal maturity is required to know and admit when one has reached the limit of one's own effectiveness. If a company cannot adapt to changing circumstances, it will either fold or drop back to a marginally surviving company.

9.4   Company Growth and Management Thresholds

The issues of resource allocation, personnel, planning, and business systems gradually increase in importance as a company progresses from start-up, through slow initial growth, to more rapid growth. It is best to acquire these resources in advance of the growth stages so that they are in place when needed. Yet, the point at which growth will outstrip management ability is not easily predicted because the threshold is different for each contractor and for each management team. There are no specific limits for the operation of a given size or type of construction enterprise. The contractor must therefore be able to identify when company growth is outpacing their managerial maturity. The best judges of this are the contractor and their management team, who are usually too busy to look for signs of growing pain or may not recognize them when they see them. Unfortunately, the easiest way for a company to identify that they have growing pain issues is to exceed them. Realizing that you have already outstripped your management ability is retrospective analysis and of little help because you discover the exposure only after the damage is done because you have already exceeded your limits before you recognize them.

9.5   Telltale Signs of Insufficient Maturity

It is critically important that contractors be aware of the risk of overextending themselves beyond their managerial threshold and to continually be on the lookout for telltale signs that change is needed. An organization that is managed by more than one person has a slight edge in that multiple observers provide a better system of checks and balances. Managers are able to observe and critique one another's methods and results, which tends to bring management shortcomings to the surface faster.

There are several telltale signs when a construction company is trying to manage more than its leadership team is actually capable of. For example, a series of internal gripes or small problems, most often concerning personnel issues, signal possible trouble. Particularly disturbing signs include:

- complaints from long-time trusted employees
- increased turnover among middle managers and field forces
- increase in owner and designer complaints
- a drop-off in the performance or response of trusted subcontractors
- increase in job site accidents
- increased absenteeism.

These factors tend to grow proportionately with the growth of the business. Yet, when several of these challenges occur at once and for no apparent reason, it is usually a sign that something is wrong. When a growing company's management team seems to always be reacting to business pressures and concerns as opposed to anticipating new issues and proactively addressing them, something is wrong. Without continuous and timely management improvements, the risk of the business failing grows exponentially as the business continues to expand. Keep in mind that we are in the second-highest failure rate industry in the nation.

When a construction organization doubles or triples in size, it is no longer the same company it was before that growth, and it usually will struggle to be successful if it maintains pre-growth management methods. Yet, changing existing management practices is extremely difficult for some contractors and for good reason. One of the more difficult changes is to evolve from the close-knit, team-oriented camaraderie of a small company into a midsized organization in which continuous close contact among all employees is no longer the norm or practical. Sometimes long-time trusted employees are unable to develop personally or professionally to function well in a larger or more complex organization. These employees may even blame management for their discomfort, and management must try to accommodate and cope with both progress and personal loyalty.

9.6 Changes in Top Management

Replacing top management personnel is also a huge challenge that many companies are unable to overcome. All too often, there is no formal succession plan in place. Most construction companies have a limited number of qualified personnel to choose from during the succession process, and many are reluctant to go outside their own ranks for new managers, even when the need arises suddenly. This reluctance is most common in closely held companies where the managers will normally have worked with and learned under the retiring leader and have no outside or independent managerial experience. Regardless of the circumstances, the new manager's managerial maturity will be tested immediately upon succession, which is a critical test of the organization's overall managerial maturity. If new leadership lacks managerial maturity in terms of competence, experience, and confidence, the success of the entire company is at risk.

Leadership transition is difficult no matter the circumstances, but in the closely held family business, which is common in the construction industry, the succession process is even more complex. The job of operating a family-owned business is often grievously complicated by friction arising from rivalries that exist between the various family members who hold positions in the business or derive income from it. Oftentimes, these barriers prove to be insurmountable and have caused companies to fail.

9.7 Delegation of Authority

An important skill needed by contractors who are managing a growing business is the ability to delegate authority. Delegation has two key benefits. First, it builds the company's managerial maturity because the contractor provides their subordinates with experience in making business decisions. And second, it lessens the workload that is placed on the contractor (Figure 9.2). However, true delegation of authority within the construction industry is a problem in and of itself. Oftentimes, the very personality traits that cause someone to want to become a contractor, such as the desire to be their own boss, make delegation difficult and, in some extreme cases, impossible.

There are far too many large construction organizations with layers of vice presidents (VPs) and managers who will not or cannot make even the smallest decisions without the contractor's input. Many contractors think this is a good system of checks and balances, but they can't be everywhere at the same time to make or approve every major and minor decision in the company. They wonder why they're working such long hours. This type of arrangement is common in the construction industry and causes missteps. It is a mistake waiting to happen. When employees are given titles and not the corresponding

Figure 9.2 Delegation of authority builds managerial authority, spreads the workload around, and is necessary to manage a growing business. *Source:* NewSaetiew/Adobe Stock Photos.

authority, they will eventually leave or, worse, mentally retire. Few will put the required effort into quality decision-making when they know they cannot make a real business decision on their own without clearing it with the boss.

9.8 Test of Delegation

The test of true delegation of authority is when the person delegated to is allowed to make a mistake. True delegation means that the person, to whom authority is delegated, is eventually allowed to make unsupervised decisions that may or may not be correct. Contractors frequently resist the suggestion that they should allow their subordinates the latitude to make a mistake because mistakes can cost the company money. Why not, they reason, allow people to make decisions in their area of authority but then require them to check in with the contractor to receive approval prior to implementation? This approach may be a valid training procedure, but trainees do not run construction companies. Eventually, decision-makers must be given the opportunity to sink-or-swim on their own abilities and live with the consequences of their own decisions.

Of course, mistakes will occur, and some will cost money, but the most important consequence is that personnel will learn from their mistakes and increase their capacity for future success. Mistakes are an inevitable component of management development. The reward is that once developed, these managers take some of the workload off the founder and higher-level top management. This process is critical to the long-term success of the organization, and it provides top management with the capability and resources to lead the company through inevitable leadership changes.

The inability to truly delegate authority has affected the capacity of numerous contractors to retain highly trained key personnel. The reason for this is that a lack of true delegation slows the advancement of middle managers, which drives away the highly qualified. Middle and top managers who are qualified to lead in the construction industry strive to become senior managers, and these types of qualified personnel will not stay long with an organization that is unwilling or unable to delegate true authority. This is even more challenging for relatives in family businesses because those in line for succession often are not allowed to practice decision-making. These family members are commonly unable to leave the business due to family pressures and are forced to endure the *no-real-authority* syndrome until their first binding decision has to be made, which is usually after they inherit the company. The lack of true delegation of authority sets companies up for failure because managers are unable to truly develop and test their management abilities and are therefore unprepared when they all of a sudden are responsible for making major decisions. Employees who are not given the authority to make decisions have no opportunity to develop good judgment.

Achieving managerial maturity doesn't just happen. The need for it must be recognized, and the skills learned or hired. Understanding the importance of real delegation of authority and managerial development in a growing construction company is the first step toward achieving it. Peter Drucker writes:

> Entrepreneurship is risky mainly because so few so-called entrepreneurs know what they're doing. They lack the methodology. They violate the well-known rules.

Case Study: Managerial Maturity

Consider a hundred-million-dollar-a-year building construction company in business for over 50 years that had several executive vice presidents (VPs), in addition to a score of other vice presidents. The growth of the company was skillfully managed by its founder for years. The founder presided over his huge organization with a strong hand and had the ability to juggle numerous tasks and responsibilities at once, including the capacity to work long hours. To the casual observer, the founder had a strong management team to back him up and carry part of the load, but upon closer scrutiny, there was very little true authority among the senior executives.

There was loyalty, plenty of hard work, and long hours, but almost all of the major decisions were made or given final approval by the founder himself. As he aged, he spent less time at the business, but given the size of the company, the important decisions could still be handled by him even if they had to be delayed a while. Because he maintained all of his personal contacts in the political and business community, the founder remained a critical element in the continued success of the company, even after he formally retired.

When the founder finally retired in his eighties, he set up a Board of Directors, which was composed entirely of insiders. In effect, the founder's top managers ran the company by committee. Some of the members had come up through the construction side of the business, while others were leaders in accounting, marketing, or human resources. There was considerable friction between the field and office factions. The office executives generally dominated the conversations at meetings and continually attempted to minimize the influence on the construction people. In several instances, they attempted to remove senior managers with field experience from the management team, as they were thought to lack sophistication.

Although completely retired and in ill health, the founder's influence was strongly felt in the boardroom of the company, even when he did not attend all the meetings. The construction personnel in senior positions had worked for the founder for many years and spoke proudly of the history and values of the company. They worked hard to continue doing business in a similar manner to what the founder had established. The office personnel on the board pushed multiple changes but were unwilling to fight for these changes with the founder present or in the background. The company remained profitable during this period, while the differences among top management grew but few changes were made.

A number of years after retirement, the founder passed away. Within months, a special board meeting was held that barely made a quorum because several construction members were at remote job sites. At this meeting, a new management committee was formed to govern the day-to-day operations of the company, and the Chief Financial Officer was elected President. Within a year, the entire construction organization was subordinated to the administrative/financial department, and a number of key construction people either quit or were fired.

The company expanded rapidly and more than doubled in size in less than three years following the founder's death. The profit picture was not very good, but management attributed this to the cost of growth and predicted it would improve over time. The rapid growth had included the introduction of new computer equipment and systems, which, when coupled with the distraction of moving to new corporate headquarters, resulted in an administrative nightmare. Data input was weeks late, and some critical data were lost. This caused a serious delay in producing a certified financial statement at year-end. Five years after the founder's death, a six-month internal interim financial statement showed an operating loss. This occurred while computer problems had delayed the prior year's certified statement, which still had not been produced. The company's surety became concerned, and bonding was restricted. Several months later, the prior year's results were determined to be a loss that amounted to in excess of 8% of sales.

The company operated for seven years after the retirement of the founder. During three of those years, the founder was alive and continued to influence the direction of the company. For those three years, performance and profits were fine and equaled historical performance. Within months of the founder's passing, everything changed. The new management team did what they called *bringing the organization up to date*. They claimed to have modernized the business. Because of the time lag from the founder's retirement, it was assumed that the new management had done well for three years. What was not recognized was that the founder's influence was felt after his retirement and until his death in a way that compelled management to continue to do business with the values, goals, and in the manner he had established.

At first glance, this business failure might appear like a change in key personnel, and even that the founder had nothing to do with the business failure. The organization did lose the founder – first to retirement, then to death. In addition, they subsequently lost several key senior construction people. While both the founder's retirement and then death were certainly a loss to the company, they cannot be considered a surprise. Everyone eventually retires and dies.

Considering the reality that every company founder will eventually be lost to any organization. This occurrence should not – and does not – have to cause the business to fail. Nor does it have to place every enterprise that goes through it at risk. This change in key personnel happened, but it was totally predictable and certainly not untimely. Therefore, the cause of the problem was not a change in key personnel but the lack of preparation for an obvious and certainly predictable event. The real cause of this business failure was the refusal or inability of the founder to select, train, and prepare his replacement in advance of his retirement. This would, of course, have had to include true *delegation of authority*.

In this case, failure was a direct result of the founder's inability – or refusal – to prepare his replacement. There was no one in the organization that had any real authority prior to the founder's retirement. Consequently, a few managers had experience running their own departments, but no one had any experience running an entire construction company. Let alone a 100-million-dollar construction company. Everyone did their job as they had been taught by the founder, who was regularly consulted if any exceptions were encountered. There was no lack of talent in this company, although the more aggressive managers didn't stay long. What the organization lacked was initiative. Everyone followed the founder, but they worked *for* him, not *with* him. There were no self-starters because they were all *jump-started* by the hard-driving founder.

The success of this company can and should be clearly attributed to the aggressive and talented founder. And its failure is also clearly attributable to the founder. What was missing was the management development needed to sustain the company's success when the founder's skills would no longer be available to the organization. It is every manager's responsibility to tutor and prepare their successors. The founder of a successful company is no exception. In this case, the founder was either unaware of this fact or was unable to successfully address it, which ultimately indicates that he had outstripped his management ability and lacked the *managerial maturity* to lead the business at the size to which he had grown it.

Some may argue in this case that building and successfully managing a business of that size proves the founder's management ability. However, if the company's continued success relied solely on the founder's continued personal involvement, then it was not truly a *viable* commercial enterprise. It is critical for business owners to recognize the need to train and prepare their successors if they intend for the business to succeed them. It is also critical for top and middle managers to train their replacements. It is management's responsibility to cause this to happen where and when needed. Top management must understand that running a business includes the use and delegation of authority. The cost of training subordinates and the time associated with teaching the delegation of authority to subordinates is nothing more than a cost of doing business.

9.9 Succession Planning

A closely held company can be defined at any given point in its development as the sum and substance of its experience, and almost all of that sum and substance is created by and resides in people. Particularly and proportionately in its key managers. Some of the sum and substance becomes institutionalized, such as attitudes toward customers, work ethics, values, and so on, because the key people have woven them into the *fabric* of the organization by example, mentoring, and training. When key people leave a closely held construction enterprise unexpectedly, as in illness, or expectedly, as in planned retirement, some of the *essence* of the organization is lost. A portion of the sum and substance can be transferred from one person to another with deliberate and diligent effort, such as customer relationships, union contacts, production and process knowledge, and the like.

However, a considerable portion of the sum and substance cannot be institutionalized, is not transferable, and, despite best efforts, is forever lost to the organization when a key person leaves. Lost to the organization are seriously important things like:

- insights into business or project risks
- the development of innovative processes, policies, and actions born out of years of company-specific experience
- masterful hiring selections
- solutions to personnel issues made with an ability to read between the lines or *hear* what is not said, using an innate talent honed over decades
- and much more.

The steps necessary to minimize the loss of sum and substance to the organization are generally referred to as a succession plan. They are:

- Discover, evaluate, and delineate exactly what the sum and substance is that is captured in a key person, well before they leave the organization.
- Determine which of these are truly institutionalized and outline steps to be taken to assure that they are, *in fact*, firmly engrained into the *fabric* of the organization.
- Identify the sum and substance that can be transferred to another person, determine who they should be transferred to, and devise a methodology for deliberately and effectively transferring them over time, and scheduling by who and by when. This usually takes two to five years, depending on how senior the person leaving is.
- Establish which sum and substance is not transferable and will be forever lost to the organization.

- Minimize the resulting impact of the above loss by recognizing that the particular talents will not be applied to certain decisions. Realizing that, in each instance, management will need to evaluate the impact the missing ingredient may have on the risks associated with the decision.
- Attempt to compensate for the increased risk by introducing additional people into collaboration before making certain decisions.
- Cultivate the same or similar talents, to the extent practical and possible in the person(s) succeeding the key person who is leaving, through a carefully developed and diligently pursued multi-year Professional Development Plan.

It is critical to accomplish the above, which takes a minimum of several years, after which it is prudent to test and verify the transfer of knowledge by having the replacement person work with/alongside the person leaving. It is best to have at least three years, and safer to have five years, for critical positions to ensure that as little as possible of corporate knowledge (sum and substance) is lost. Unfortunately, there is often no notice when a middle or top manager leaves because it is often unexpected. For key, very senior management departures, five years of preparation is appropriate.

When there is uncertainty about when a person may leave, such as an open-ended intention to retire in a non-specific length of time or depending on health issues, it is prudent to plan for the shortest time. If the person stays longer, all the better. The more senior the person, or the greater the amount of sum and substance presumed to be lost, the greater the effort and resources that should be applied to the succession planning.

9.10 Succession Issues

A candidate to replace a senior position in a construction company is generally selected for their experience, including responsible senior management positions within the organization or at a similar company. An attempt must be made to determine if a candidate has the necessary experience to develop the skills and also possesses the talents necessary to maximize the potential for success in the new position.

Necessary Skills
Skills are attempted to be measured in the broad areas of technical, leadership, business, finance, and marketing. Below is a broad list of skills as an example. Each organization will have its own list of skills and attributes they require for a particular senior position. An example of minimum skill sets for each heading is:

Technical – An understanding of scaffolding systems, equipment utilization/management, concrete forming systems, characteristics of concrete and mix design, planning and scheduling, productivity measurement principles, blueprint reading, estimating takeoff/pricing, field safety, and OSHA regulations. Exposure to soil mechanics, strength of materials, and basic physics.

Leadership – Understanding of leadership styles, leadership principles, principles of motivation, the difference between management and leadership, the use of authority and power, impacts of change, maintenance of trust coalitions, and significance of vision.

Business/finance – Understanding of basic accounting, bank credit granting principles, how surety works, the importance of cash flow, business planning methods, basic contract law, contract administration, construction cost accounting, and alternative dispute settlement techniques. Exposure to basic statistics, basic risk analysis, insurance principles, business ethics, computer technology, and Generally Accepted Accounting Principles.

Marketing – Understanding of basic marketing principles, public relations, principles of selling, marketing intangibles, and the importance of understanding and measuring the competition. Exposure to market analysis principles, types and purposes of advertising, and concepts of consumer behavior.

Assumptions (This list is an example. Each company's list should be specific to the organization and the position in question.)

Candidate has a mastery of written and oral communication skills, is computer literate, possesses good people skills, and has been educated in basic accounting methods and principles, micro and macroeconomics, business law, basic psychology, report writing, business research methods, management theory, computers in business, and personnel management.

Learning Methods (Particularly if there are shortcoming in the required skills.)

Skills can be learned and/or enhanced through on-the-job training, in coursework, or they can be self-taught, primarily through reading. While some believe that not every skill can be learned through experience, it is sufficient to state that some are simply learned more efficiently or effectively through one method rather than another. It is believed that at least some coursework (typically evenings) may be necessary, but attempts will be made to minimize that method. Alternatives of possibly less efficient learning methods may be substituted where possible.

Experience, by definition, can only be gained by doing or on-the-job training. However, carefully designed field and office experiences, vigorously perused, may shorten the time significantly with little sacrifice in learning.

Minimum Experience

It is extremely difficult to lead others in an organization without the ability to do what you direct others to do. Construction industry field people have a high regard for ability and seldom perform well when they perceive the person directing them as less skilled at the task than they are. Most construction industry mid- and top-level managers have field experience and have similar expectations of their superiors. Some consider it the nature of the business. Perhaps more importantly, there is great exposure in requiring people you supervise to perform, absent your ability to know, first-hand, whether they are functioning correctly or efficiently.

Competency

Examples of minimum competency tests are to demonstrate the ability to:

- Build a midsize typical company project (+/− $0.5 million) as the superintendent generating a reasonable profit.
- Estimate, price, and capture (or have the right bid) a midsize typical company project ($100 K–$1 million) performing under normal supervision.

- Project manage a midsize typical company project ($100–500 K) on time on budget.
- The operations manager successfully supervises all facets of multiple projects, including cost accounting, with Project Managers reporting directly to President/CEO.
- The operations manager, working in coordination with the VP in charge of finance, demonstrates an understanding of the company's accounting systems, ability to interpret financial reports, and present financial data clearly and concisely to the satisfaction of company's bank and surety representatives.

Education and Training

Certain knowledge is necessary for learning or perfecting of skills. Not easily learned from the above experiences some may have been learned in coursework. The following subjects are very brief examples of particular skills that an organization may consider necessary. If these skills have not already been mastered through experience, they will need to be learned in coursework, coaching, or on-the-job training. Some may be able to be self-taught.

- Scaffolding systems
- Equipment utilization/management
- Concrete forming systems
- Characteristics of concrete/mix design
- Planning and scheduling
- Productivity measurement principles
- OSHA regulations
- Soil mechanics
- Strength of materials
- Basic physics
- Estimating takeoff/pricing
- Difference between management and leadership
- Use of authority and power
- Impacts of change
- Maintenance of trust coalitions
- Significance of vision
- Bank credit granting principles
- How surety works
- Importance of cash flow
- Business planning methods
- Marketing intangibles
- Understanding and measuring the competition.

Much of the above can be learned from books, experience, evening courses, etc. Many topics are available through industry seminars, and some through discussions with industry professionals inside or outside of the company. The subjects, methods, and timetable can be flexible, depending on the responsibilities of the position in question.

9.11   Measuring Succession Progress

After several months in a new senior position, it is appropriate to measure progress. One method is self-evaluation, with an evaluation by management using the same short questionnaire. This can also be helpful in testing future candidates for promotion. Below is an example of a short questionnaire that readers are invited to alter as they see fit, including a brief introduction to the users.

CORPORATION SUCCESSION PLAN

Candidate's name _______________________________   Date ______________

PROGRESS QUESTIONNAIRE

INTRODUCTION

The company is investing heavily in assisting you to accelerate your growth to fill the shoes of those who will be advancing or retiring, so please take the time to think through these questions carefully. Your thoughts and feelings in each area are critical to the evaluation process and will form the basis for you and management to discuss changes or improvements to your professional development. Please answer the following questions on a scale of 1–10, with 1 being little progress, 5 being some progress, and 10 being excellent progress. Feel free to add comments that you feel apply and return the completed form by _________.

Progress in being introduced to client contacts

1   2   3   4   5   6   7   8   9   10

Comments: ____

Progress in establishing relationships with new clients

1   2   3   4   5   6   7   8   9   10

Comments: ____

Progress in managing and improving existing client relationships

1   2   3   4   5   6   7   8   9   10

Comments: ____

Progress in being introduced to engineers

1   2   3   4   5   6   7   8   9   10

Comments: ____

Progress in establishing and improving relationships with engineers

1   2   3   4   5   6   7   8   9   10

Comments: ____

Progress in being introduced to vendors

 1 2 3 4 5 6 7 8 9 10

 Comments: _______________________________________

Progress in establishing and improving relationships with vendors

 1 2 3 4 5 6 7 8 9 10

 Comments: _______________________________________

Progress in attendance at association meetings, golf outings, vendor events, etc.

 1 2 3 4 5 6 7 8 9 10

 Comments: _______________________________________

Progress in being introduced to union contacts

 1 2 3 4 5 6 7 8 9 10

 Comments: _______________________________________

Progress in establishing and improving relationships with union contacts

 1 2 3 4 5 6 7 8 9 10

 Comments: _______________________________________

Progress in understanding how to assign superintendents and what to consider

 1 2 3 4 5 6 7 8 9 10

 Comments: _______________________________________

Progress in understanding mobilizing and moving equipment effectively

 1 2 3 4 5 6 7 8 9 10

 Comments: _______________________________________

Progress in participation in regular ride-along with senior managers

 1 2 3 4 5 6 7 8 9 10

 Comments: _______________________________________

Progress in interactions in field decisions to gain understanding of management's thinking

 1 2 3 4 5 6 7 8 9 10

 Comments: _______________________________________

Progress in training and Q&A sessions with supervisors

 1 2 3 4 5 6 7 8 9 10

 Comments: _______________________________________

Progress in being included in interactions with candidates for employment, interviews, and hiring

 1 2 3 4 5 6 7 8 9 10

 Comments: _______________________________________

Progress in being instructed in production and process knowledge

  1   2   3   4   5   6   7   8   9   10

    Comments: ____

Progress in being instructed in project risks

  1   2   3   4   5   6   7   8   9   10

    Comments: ____

Progress in being instructed in hiring selections and solutions to personnel issues

  1   2   3   4   5   6   7   8   9   10

    Comments: ____

Progress in being instructed in pre-bid evaluation, pricing, and risk assessment

  1   2   3   4   5   6   7   8   9   10

    Comments: ____

Progress in understanding how to mentor superintendents

  1   2   3   4   5   6   7   8   9   10

    Comments: ____

Progress in understanding issues of micromanaging superintendents or projects

  1   2   3   4   5   6   7   8   9   10

    Comments: ____

Progress in managing effective team building

  1   2   3   4   5   6   7   8   9   10

    Comments: ____

Progress in becoming a safety champion

  1   2   3   4   5   6   7   8   9   10

    Comments: ____

Progress in mastering negotiation

  1   2   3   4   5   6   7   8   9   10

    Comments: ____

Growth toward understanding a company-wide view of your area of responsibility

  1   2   3   4   5   6   7   8   9   10

    Comments: ____

Growth toward understanding the skills required to be an outstanding manager

  1   2   3   4   5   6   7   8   9   10

    Comments: ____

Understanding of areas you need to improve in – list in comments below

| 1 | 2 | 3 | 4 | 5 | 6 | 7 | 8 | 9 | 10 |

Comments: ____

What subjects or activities would you like included in?

Comments: ____

____

9.12 Summary

Changes in key personnel are clearly an exposure for the small and midsize construction organization. The reality is that a successful construction operation is made up of knowledge, skills, and processes, which are housed in key personnel. As people leave an organization, they take their knowledge with them, and the organization is changed. Understanding this obvious, but seldom considered, reality is critical to the continuing success of a construction operation.

There are numerous successful construction companies that will be at risk when the founder, current leader, or senior managers are no longer with the firm because of a lack of realistic successor training. They will not all fail. The new leadership may have the innate ability to do the job. However, most companies are at risk because they face the transition without the benefit of a trained replacement. A trained successor is someone who has been given the authority to practice running the company, division, group, or part of it and has demonstrated their ability and honed their skills on the proving ground of real authority.

Many contractors claim they already have too much to do in managing a growing construction business to be able to deal with time-consuming successor training. This attitude is just another nail in the coffin of growth, because if an enterprise cannot control its growth long enough to train people, they are putting the company at unnecessary risk.

Other contractors may say, *let them learn it the way I did*. This may work for a construction start-up where mistakes are made on a smaller scale and are more easily recovered from. Learning by doing or by trial and error is accelerated by the rapid-fire issues that confront the smaller or start-up organization, but assuming responsibility for an ongoing construction operation can be like trying to jump onto a moving train.

Learn it the way I did does not make sense when taking over an ongoing business without realistic training anymore that it would trying to operate a moving train without training. It is unrealistic to think a newcomer can operate a company as well as the founder, who has practiced on it for 30 years to get it to its current and largest size. And it certainly does not mean it will be easy to take over a company that even the founder struggles to manage. Anyone who uses this sink-or-swim training method should not be surprised at its randomly selective results.

10

Understanding Construction Accounting

Accounting is the language of business, and information is communicated to company owners and managers through numerous reports and through financial statements prepared, at least annually, by independent auditors. This topic is critical and may be uncomfortable for some industry professionals because owners and senior managers often express limited interest in accounting processes. Let's begin with a clarification: this chapter is not about how to conduct accounting functions. These functions are performed by accounting professionals in a construction firm's accounting department and by independent Certified Public Accountants (CPAs) who audit the accounting department's work. The purpose of this chapter is to provide nonaccounting construction professionals with insights and an amplified understanding of the accounting information they are exposed to in the course of their work.

As we have previously mentioned, accounting is not a language that the average construction professional is fluent in. For most people in the construction industry, accounting is more like a foreign language. This situation is understandable. Construction professionals are highly trained in the technology of constructing and the built environment, and just as with surgeons and pilots, construction professionals cannot be expected to have studied other professional disciplines.

The lack of training in accounting becomes a problem that is seldom recognized when a contractor's vocation grows into a business. Because construction is so highly technical, many practitioners do not see their vocation as a business. Some members of the industry have demonstrated an irrational level of disregard, even disdain, for business elements such as accounting, economics, finance, statistics, marketing, business management, human resources, business law, and psychology. We have been told too often that what matters is *getting* the project built. That perspective is frightening because what really matters is *making* a profit.

Most industry practitioners understand the importance of making a profit but believe that as long as they are profitable, someone else can count the money and keep track of the details. After all, how hard can that be? We have also heard contractors say, *We come by our accounting skills naturally. We have good instincts.* Our response to that is that if a contractor does not know much about economics, statistics, and the other business elements, how can he or she possibly know which of those things are required at any point in time? More significantly, how can the contractor be even remotely qualified to supervise those things? If we are honest with ourselves, most contractors are not qualified accountants, and in reality just accept the accounting outputs they are given. If the outside accountant says the company made $451,562.17 in profit,

The Business of Construction Contracting: Schleifer's Guide to Financial Success, First Edition. Thomas C. Schleifer and Aaron B. Cohen.
© 2025 John Wiley & Sons, Inc. Published 2025 by John Wiley & Sons, Inc.

the contractor has to assume that amount is correct. No one seems to wonder: If the profit calculations include the estimated cost to complete work in progress (WIP), how can such financial statements that are based on estimates be considered accurate to two decimal places?

Our purpose here is not to lament the complexity of construction accounting, which when misunderstood, contributes to construction businesses failing. Rather, our purpose is to bring these issues to light so that misunderstandings and business failures can be averted. Some people may wonder why if this topic is so important, it is not addressed more commonly? One reason is that decades ago, accounting knowledge did not matter as much because the construction industry was so profitable that any shortcomings in accounting processes were covered by continual positive cash flow. However, this situation has changed. Profit margins have been declining for quite some time.

Company owners need to stop assuming that since their organizations are completing bigger projects, hiring more people, paying bigger bills, and cashing bigger checks, everything is okay. Contractors need to stop attributing any negative metrics to temporary growing pains. Effective accounting can identify whether a firm is financially healthy, but the contractor has to be looking at the right things. Managing the business side of construction, which includes accounting, finance, and economic processes, is as important as producing the work. Some people argue very effectively that the business side of construction has become more important than producing the work. This possibility prompts a question: Is a firm successful when its audited financial statements indicate that it generated a modest profit, and then sometime later it is discovered that the firm did not generate a profit at all?

10.1 Annual Financial Statements

The financial statements that a construction company must have prepared annually by a third party include a balance sheet, an income statement, and a statement of cash flow. The processes and procedures used in preparing financial statements for construction enterprises in the United States must align with Generally Accepted Accounting Principles (GAAP). The company recordkeeping is managed and recorded by the construction firm's staff, and while following GAAP rules, there is some flexibility in reporting transactions. For example, an event can result in different measurements of income based on various factors, such as the firm's estimate of how much work has been completed in an accounting period and the firm's estimate of the cost to complete the remaining work. The measurement of profits on uncompleted contracts is complicated and relies on estimates that are difficult to verify, which presents the potential for error. It also presents the potential for manipulation if someone feels the need to do so.

10.2 Internal and External Financial Statements

For many construction firms, the main reason to prepare financial statements is to satisfy the requirements of various external parties that hold a vested interest in the financial health of the firm. Examples include bankers, insurance brokers, and bonding companies, all of which require independently prepared financial statements as a condition of continued credit granting. This financial information is used for activities such as assessing creditworthiness to maintain a working line of credit or to set bond limits.

Financial statements are primarily used to indicate the financial health of a company, but most companies produce some statements for their own use. Producing internal financial reports for use in running the business is part of what is commonly called *managerial accounting*. Income statements help contractors understand their performance, such as how much money they spent on the direct costs of construction over the past month or year. The balance sheet is a convenient way of providing a consolidated view of the cash a company has in its bank accounts, how much money the company owes (payables), and how much money the company is owed (receivables). These internally produced statements are useful in the day-to-day management of a company but are subject to certain inaccuracies. For example, the sales that were included in a statement may reflect a change order that has yet to be approved. Alternatively, the expenses recorded may not include an invoice from a subcontractor that has not yet been entered into the accounting system. In some cases, the invoice may still be sitting on the project manager's (PM's) desk because he or she is disputing whether the subcontractor completed a task that the invoice claims to have been completed. Any inaccuracies can obviously render the financial reports less useful in managing the business.

Additionally, the financial reporting process and format produced for managerial accounting purposes often differ from company to company because many companies conduct business differently. Differences in the way companies are organized and conduct business typically influence the various financial metrics a company may deem critical to its success. Focusing on these metrics can result in unintentional bias, affecting the manner in which the company produces the statements and interprets the statement's information.

Because internally prepared statements have potential weaknesses, external stakeholders require companies to provide independently prepared financial statements. These statements are prepared by an independent auditor, who follows a consistent process in order to ensure that the statements have been prepared in accordance with GAAP, are free of material errors, and are as accurate as possible. These statements are accompanied by a declaration that, in the auditor's opinion, the financial statements represent a true and fair view of the company's financial performance and position. Therefore, audited financial statements provide bonding companies and banks with a certain level of assurance regarding the accuracy of the financial information they are using to make financial determinations.

Audited financial statements may differ from internally prepared statements. For example, as previously mentioned, internal statements may include in sales totals the value of an unapproved change order because the contractor is confident that the majority of the change order will be approved. External auditors usually include only signed change orders because until the change order is approved and signed, it is not officially or contractually included in the contract in question.

Annual audited financial statements take some time to prepare and may not be available until months after the fiscal year ends. When the statements are finally available, contractors seem to pay limited attention to the information contained in the statements. One reason is the belief among some contractors that accounting is a cost of doing business that does little to enhance a company's operations, and they have little interest in information that is months old. We strongly believe that contractors need to understand how financial reports can either contribute to or impair the success of a business. Contractors must recognize how externally prepared financial statements, based on internally furnished information, are used by external stakeholders. These stakeholders use the statements to understand a company's financial health and then make important decisions that will affect the company's ability to borrow money and take on projects in the future (Figure 10.1).

Figure 10.1 Accounting is the language of business, and financial statements are used to communicate the health and performance of the company to internal and external stakeholders. *Source:* wirojsid/Adobe Stock Photos.

10.3 Management Accounting Inputs

A construction company handles a lot of money that is not the company's because it is owed to subcontractors and suppliers, and contractors get to keep only the profit, which many people agree is too small for the risk involved in the construction industry. However, no one seems able to convince industry employees or the general public that construction profits are small. Most medium-sized and larger general contracting firms would be happy to keep 2–5% of sales, and self-performing specialty contractors work hard to earn 6–10%.

Because the work is complicated, materials are expensive, and schedules are often erratic, calculating the current profit on work in progress is difficult at best, and in some stages of the work, is almost impossible. Further complicating the matter, the amount of work put in place may not be the amount of work that is accepted by the designers and/or that can be invoiced for. Some work may require rework later or accrue other costs after the financial statements have been prepared and relied upon.

The primary risk regarding the potential inaccuracy of financial information is the profitability of work in progress. Calculating the profitability of this work requires accurately and timely inputting of information known only to the people actually engaged in the work. These people are in the field, whereas the accountants and bookkeepers are in the office.

Owners and senior managers who come up through the ranks or who have earned construction or engineering degrees may never have had a basic accounting course. They probably understand basic mathematics but have never had to deal with debits and credits. If a contractor does not participate in the accounting function of the business or has a limited interest in accounting, the bookkeepers and accountants are left to collect work-in-progress information and develop reports and statements without the help of operations personnel who can evaluate, from everyday knowledge, whether the information makes sense. It bears repeating that many contractor-owners of some smaller companies have so little interest in the accounting side of the business that they dismiss the value of financial statements as a tool to more effectively manage the business.

10.4 Methods of Accounting

We never tire of stating that the primary reason for being in business is to make a profit. If someone happens to be in the construction business for another reason, he or she will not be able to stay long without making a profit or at least breaking even. Profiting on a job can be difficult. On some jobs, the difference between earning a profit and losing money is extremely small and could depend on just a couple of good or bad breaks. The only thing harder than profiting on a job is accounting for the profit. Indeed, one of the most difficult and misunderstood processes in the construction business is capturing all the data accurately and in a timely fashion, and using it properly to track interim profit or loss and then final profit or loss.

Construction accounting is only easy if the job lasts about a week, if everyone bills properly and on time, if the contractor does not forget the fringe benefits or the cost of company-owned equipment, and if the contractor eventually collects retainage and remembers to back in the cost of callbacks and guarantee work. Accounting gets more complicated when a project lasts longer than a week. Numerous accounting systems and processes are required to determine whether the organization is making or losing money and by how much. As with many other aspects of the construction business, accounting for construction work in progress differs from accounting for work in other industries. The accounting methods used effectively in many other industries often fall short when applied to the construction business, especially as a construction company continues to grow.

GAAP rules govern how financial statements are prepared, but because the accounting rules for construction differ from those in the majority of other industries, it is important to have in-house accountants and independent CPAs who understand the differences. This knowledge may be less important for start-up companies, because any losses are immediately brought to light by cash flow difficulties. This knowledge is vital when a contractor grows, particularly rapidly. These companies quickly outgrow the accounting procedures that worked fine for them when they were small. So, the best option is for companies to implement, from the very beginning, the construction accounting practices that will enable them to understand the accounting process and accurately measure their progress and profitability.

10.5 Accuracy of Accounting

It is necessary to account for all of the money that passes through an organization in order to know how much the organization gets to keep. Whether or not the contractor and top managers have accounting experience, they need to participate in the accounting procedures. The contractor and managers do not need to do the actual accounting work but must understand the systems to ensure the numbers make sense from month to month. To that end, the owner and managers must be involved in the accounting function each month. (Most are not, but that does not alter the reality that they should be.) In small organizations, the accounting systems must make sense to nonaccounting managers and meet the organization's need for a tool to run the business. For midsize and larger firms, the systems may be too complicated for nonaccounting managers to understand; however, the system's reports need to make sense to nonaccountants. In either case, the results must be accurate, and it is the CEO's responsibility to make sure they are accurate.

A construction company's bookkeepers and accountants obviously lack firsthand knowledge of what is occurring in the field and, therefore, must rely on information from PMs and operations personnel. When top managers do not regularly scrutinize the accounting results and simply accept them at face value, inaccuracies go unnoticed. We have seen overhead costs reported and accepted that were 70% of the previous month's overhead costs, and no one sensed that there must be an error. We have also seen sales reported improperly, with no one realizing that billing did not go out on one of the largest projects that month.

A contractor who continually monitors the numbers side of the business knows approximately what the month's volume and costs are before the accounting department does – and rightfully so. The bookkeepers and accountants can only add up that which is given to them in the normal course of business and put these figures into the proper categories to develop accounting reports. But if vendors or subcontractors neglect to invoice for their products or services in a timely manner, only the field managers, PMs, or contractors know that the material was received or that the work was done and that a liability was incurred. At the risk of repeating the obvious, the accounting department is responsible for accurately recording and reporting the information received, but it is management's responsibility to ensure that the accounting department receives correct information in a timely manner. This responsibility includes establishing processes for that to happen and the supervision of the processes to see that they are continuously delivering the correct information.

Some contractors contend that as their businesses get larger, management can no longer pay close attention to the numbers each month. But the numbers *are* what matter. To manage properly, the contractor and managers must regularly look at the accounting reports and know whether the information is logical. Are they certain that all the costs incurred are included in the report? Does the report show a 20% gross profit when they know that the profit is actually lower? Ensuring that the reports make sense may mean acquiring better systems and disciplining the use. In a small company, that may mean hiring a controller instead of a bookkeeper. In a larger firm, that may mean hiring a chief financial officer (CFO) (Figure 10.2) instead of a controller. Critical changes such as these need to be made as the size of the company increases. If management does not have the time or training to manage the accounting function, they should consider instituting a board of directors or advisors to serve as a safety valve. The best option is to choose board members from outside of the company. In our experience, independent board members always seem to know whether the numbers make sense because they are far enough away from the trees to see the forest. The bottom line is that a contractor cannot relinquish his or her responsibility for ensuring the accuracy of the firm's financial reporting.

Figure 10.2 The Chief Financial Officer (CFO) is ultimately responsible for the accuracy of the companies' accounting functions. *Source:* moodboard/Adobe Stock Photos.

10.6 Percentage of Work Completed

The state-of-the-art in construction accounting has advanced over the years but continues to be an area of great concern. As we previously indicated, a unique aspect of construction accounting is the rules used to account for work in progress. When a project is partly complete, GAAP requires that financial statements report the portions of revenue that have been earned, and these earnings are estimated based on the amount of work the company claims to have been completed. The primary problem with basing financial information on a percentage of completion is that an independent auditor will have great difficulty verifying how complete each project actually was at the time that the information was reported. Some outside accounting firms have said, *We have our own way of testing those numbers.* This statement is interesting in light of the fact that contractors themselves cannot verify the numbers they provide to the accountants concerning uncompleted contracts. It is close to impossible for a construction professional to accurately determine whether an ongoing project is 39, 40, or 41% complete at the end of a given month. Nevertheless, almost every accounting report includes a completion percentage, such as 39, 40, or 41% for each project. Are no projects ever 39.5% complete? Most accounting firms that specialize in construction have developed processes that test the numbers that clients provide. The difficulty is that the project information goes through a number of individuals' hands, from field to office, and those who provide the numbers describe them as estimates. The word *estimate* does not imply *exact*.

There would be some relief if the completion-percentage estimates for all projects in progress would average each other out (with some being estimated high and some estimated low). Unfortunately, this wishful thinking defies logic. There are plenty of reasons to be accurate, but there is also an inclination to be very careful not to underestimate. The employees who estimate the percentage complete would not logically want to jeopardize their future performance assessments and bonuses by guessing low. In our analysis of tens of thousands of projects reported in audited financial statements, we found that the final percentages of completion compared against the prior period's interim percentage of completion almost always decreased over time, meaning that the mid-project estimates were overstated. If any percentage completions are overestimated, then the profits have been overstated. In these situations, some of the profits stated in the mid-project financial statements never did exist; they were a result of inaccurate estimates of existing conditions at that point in time.

Mid-project overestimates have a huge impact on the annual financial statements for a construction company because some projects do not start and end during one fiscal year. Consequently, a portion – in many cases, a large portion – of the annual profits shown on an audited annual financial statement are based on the estimated completion percentages of works in process. Some people may assume that these estimates were verified by independent auditors. However, our research shows that project profits were consistently overstated in thousands of construction companies' financial statements, which means that the estimates were not verified or that the verification was inaccurate. We are not suggesting that the audits were incorrect or that the construction firms deliberately overstated their profitability. We are simply explaining that audited financial statements are based on estimates that are ripe for human error, particularly when the people involved in the process are motivated to ensure the numbers are accurate but are also motivated to make sure that are not understated.

According to GAAP, the estimated profit on work in progress is to be reported as earned income. If the amount of work completed is estimated to be higher than it actually is, the cost to complete the remaining work will be underestimated. As a result, financial statements will indicate that more profit was earned during the period than was actually earned. If a company is growing, it will have more and more work in progress each year, and if the percentage of completion estimates are usually slightly on the high side, previous errors in percentage-complete estimates will be covered by new projects. If the percentage-complete estimates are exaggerated one year, the overestimates may very well go unnoticed as long as there is more work in progress the following year.

Some might claim that the mid-project estimates are not of consequence, because when the projects are completed and closed out, the totals are reported. Nevertheless, managers need to know where the organization is in real time, including whether the firm is earning or losing money on the present work in progress and how much. One of the greatest risks a contractor faces is to continue to work on a project in a manner that is not making a profit because they believe it is making a profit, because the accounting reports say it is profitable.

10.7 Work in Progress

GAAP requires that construction companies estimate the percentage of project completion so that a company can record the profit or loss for the portion of work completed in any accounting period (e.g. monthly, quarterly, or annually). If a project is 40% complete at the end of an accounting period, the company can record 40% of the total expected profit from that project as earned, regardless of whether the amount has been invoiced. This method requires calculating the total cost of the work to date and estimating the cost to complete the work. Auditors review these estimates when completing certified financial statements and may even attempt to check the estimates against costs of recently completed work. However, most auditors would agree that they cannot guarantee the estimates' accuracy, and GAAP does not hold auditors responsible for the estimates' accuracy.

Before a construction project begins, the cost of the work is estimated. The exact cost will of course not be known until the project is completed. When a project is partially complete, the total cost of the finished project remains unknown, making it hard to accurately predict the total cost to complete the work and therefore the expected profit. The contractor, not the independent auditor, is in charge of estimating the completion cost. Therefore, estimated amounts are introduced into what might otherwise be considered precise accounting data. This imprecise data have led to financial statements that are less precise than the financial statements in other industries. In the construction industry, the potential for error in accounting for work in progress is significant because revenue from work in progress is based on an estimate. There are hardly ever exact or even similar projects to compare costs with. And if there were similar projects to compare to, the completed project(s) would be from a prior time and require adjustments for current and future unknown inflation.

To help readers better understand how work in progress affects on the calculated profit in a construction company, let's compare how work in progress is calculated in a manufacturing company. In a manufacturing company, the value of ongoing projects is a small part of the company's total revenue and profit for an accounting period. For example, automobile manufacturers boast of producing a vehicle in less than a day. Even custom-made heavy machinery seldom has a production cycle of more than a few months. Therefore, work in progress will represent only a small portion of a manufacturing company's total revenue at the close of an accounting period. Consequently, the accuracy of the total revenue recorded is not materially affected by work in progress. Even when the manufacturing cycle is months long, the exact cost of similar completed products is known, and there is no challenge in calculating and comparing the exact costs associated with the work in progress. In construction, projects last much longer, often spanning multiple fiscal years. Determining the profit from work in progress is more complex and error-prone because no two jobs are the same and because accurately reporting the percentage of work completed to date relies on estimated values.

Case Study: Profit Margin Decline

Consider the following case study. A construction company has been in business for 15 years and has grown to a volume of $70 million. When the company started, it encountered various difficulties and struggled to establish itself in the market. Eight years later, sales were $10 million, and the business was stable. The market was booming, and the organization set out on a growth streak, almost doubling its size every two years or so. The size and number of projects consistently increased, and there were more and larger projects in progress all the time. When the company entered its 15th year of business, its audited financial statements indicated the company had modest profits and a fairly small net worth for its size. The firm was heavily in debt, largely because the firm had financed its growth and was in a weak cash flow position.

After the company began its growth phase, each financial statement showed not only more new projects each year but also more work in progress from previous periods. Each year, new work is added to the work in progress to the point that the portion of sales represented by work in progress consistently exceeded the portion of sales from the completed projects. Our study determined that over the prior years, the number of completed projects that lost money was increasing. However, the completed projects were generally smaller than the current work in progress, which was ever-increasing in size, so the losses were more than compensated for by the profits from the new work in progress.

Because the company was growing so quickly, neither field personnel nor managers had time to notice that a number of projects that were profitable at their mid-project stage had lost money by project completion. If managers had noticed the change from profit to loss on projects, they apparently were not concerned, given the firm's continuing overall profitability, as shown in the financial statements.

Finally, the firm's cash ran out, and a subsequent analysis of all past and existing projects showed some startling information. The overall picture was a disaster, but what was most interesting was that out of all the projects completed during the past several years, more than 50% had lost money. These losses were covered by the overestimated completion percentages of the always-larger works in progress. Managers did not do anything to correct the losing projects because the managers did not know the losses were occurring until the jobs were finished. Ultimately, the company failed because the profit and cash flow on work in progress failed to outpace the losses incurred upon completion of losing projects.

10.8 How to Account for Work in Progress

Because of the length of time it takes to complete a construction project from start to finish the amount of work in progress in any given financial statement is considerable. It is common for a construction company to have 50% or more of total revenue in any given fiscal year represented by work in progress. This is unique to this industry. For example, in manufacturing, the majority of products will have been produced, warehoused, and even sold during a fiscal year, so costs and profitability are verifiable and known. The auto industry produces literally hundreds of vehicles per month and in some cases per week, so costs are able to be measured accurately. The manufacture of smaller products is even faster. Because completed projects in construction may represent half or less of revenue in a year and because projects commonly differ from each other, there are little or no cost comparisons or verification based on completed projects. It is difficult to estimate the completion percentage of a construction project in progress and even more difficult to measure the costs incurred to date. If these elements are estimating the total revenue on a financial statement, it must be an estimate.

The entire process of construction accounting and auditors' creation of certified financial statements depends heavily on the correctness of four inputs: the contract price, the direct costs to date, the estimated cost to complete the remaining work, and the amount billed to date. These inputs significantly affect the precision and accuracy of financial statements, whether internally or externally prepared (Figure 10.3). The following sections explain these inputs and summarize the challenges encountered in ensuring the accuracy of these inputs.

Figure 10.3 Field management is ultimately responsible for the accuracy of the inputs that are used in producing a company's financial statements. *Source:* ArLawKa/Adobe Stock Photos.

10.8.1 Contract Price

After entering the name of the project, the first input on the WIP schedule is the contract price, which is the total value of the contract at the end of the accounting period in question. This value includes the original contract amount and any signed change orders to date. Auditors can verify this input during an audit by examining source documents. You would assume this would be a straightforward number, but like so much in construction accounting, it can be controversial. A strict interpretation of this process would suggest that this value would not include any unsigned change orders, change orders in progress, or disputed change orders. However, it is not uncommon for some or all of the costs associated with these pending change orders to be included in the recorded direct-costs-to-date. In actuality, change order work often proceeds while the change order paperwork is still in process. For this reason, internal accounting often includes the unsigned change order amounts in the contract amount to date. In the authors' experience, the unsigned change order amounts often make it from the contractor's internal accounting into the auditor's financial statement accounting. The exposure, of course, is if the change order is subsequently not approved and signed, or the amount is altered, the prior financial statement was inaccurate.

10.8.2 Direct Costs to Date

The direct-costs-to-date entry includes the costs recorded to date in the contractor's accounting system. These costs include payroll and payroll-associated costs, vendor and subcontractor invoices that have been received and approved, and miscellaneous costs and expenses charged to the project. The amount input in this entry will be inaccurate if invoices have not yet been received, or invoices have not yet been approved and recorded, or are in dispute. In our investigative work for sureties, we always found project costs that were not recorded in a timely manner. It is challenging at best for independent accountants to audit and confirm the accuracy of direct costs to date, even months after the end of the fiscal year. If any costs incurred do not make it into accounts payable before the books are closed for the accounting period, profit will be overstated. The missing costs will of course eventually be recorded in a subsequent accounting period but the profit calculation for the period in question was inaccurate.

10.8.3 Estimated Cost to Complete Remaining Work

The estimated cost to complete the remaining work is a critical input and one of the most challenging estimates to calculate in construction accounting. The amount of work yet to be completed is, of course, calculated by determining the amount of work already completed. As we have already discussed, estimating the work already completed is extremely difficult, even for those working on the project daily. Picture a large project with dozens even hundreds of tradespeople working at the site. At the end of the last day of a month, a knowledgeable superintendent or PM cannot just walk-around the site and determine with any certainty exactly how much work has been completed in each of hundreds of categories.

If the percentage of work completed cannot be determined exactly, the exact percentage of work left to complete cannot be known either. Nevertheless, the quantity of the work yet to be completed must be estimated, and so must the cost of completing that work. The amount is typically calculated using the following formula: original estimated total cost *minus (−)* cost to date *equals (=)* cost to complete. We have already pointed out the potential for unrecorded costs being missing within this calculation and that cost to complete is an estimate, not a hard number. It is extremely difficult for an auditor to confirm the accuracy of these numbers until the work is completed and accepted and all associated costs are recorded.

10.8.4 Amount Billed to Date

The amount billed to date is the total amount the contractor has invoiced its clients for the work that has been put in place. A weakness in this amount is that much if not all of the work in place may not have been accepted by the designer and/or the project owner. Interim acceptance is of limited value because it has never been found to be legally binding on the owner or designer. While these amounts are used in a monthly financial statement, the designer may not subsequently approve the amount billed, and the owner may not necessarily pay that amount. Therefore, there is the potential for this amount to be overstated and/or to be different than the amount that is actually received in payment. This entry can be audited mathematically, but it is challenging to confirm the exact value until payment is actually received and recorded.

10.8.5 Notes About Accounting for Work in Progress

It can be more difficult to accurately determine the amount of work remaining on a project than it was to estimate the cost of the work originally when preparing the project bid. The costs incurred on the work already completed are commonly used when estimating the cost of completing the remainder of the work. However, considering the costs already incurred is problematic because these costs may not be representative of the costs for the remaining work. Therefore, considering the current costs can lead to inaccuracies. A simplistic example is that the cost of placing the brick veneer on the first floor of a high-rise building is always less than the cost of placing the brick veneer on the upper floors.

Some professionals assume that the estimated cost to complete the work is an estimate that estimators or PMs prepare after evaluating the remaining work. While this process would be ideal, it is not what occurs in most cases. For the most part, the cost to complete the work is calculated by subtracting the cost to date from the original estimated cost for the entire project. However, if the percentage of the work completed is not accurate, then the cost to complete is not accurate. As described previously, the percentage of completion tends to be overstated, which understates the cost to complete. The inaccuracy of these data mean the accuracy of the profits calculated from these data are questionable; too often, profit tends to be overstated.

Additionally, the direct costs to date tend to be understated because of the difficulty in receiving and timely recording accurate costs from vendors and subcontractors. The ability of the accounting department to accurately document these costs hinges on the timely receipt of the relevant information from vendors and an internal rapid approval process of vendor invoices. If an accounting department prefers to wait for the PMs' approval of invoices before recording subcontractor and supplier expenses, then the direct-cost-to-date entry is subjected to further underreporting of the actual cost of the work.

The WIP schedule includes calculations for overbilling and underbilling in an attempt to unravel some of the challenges we have previously discussed. We contend that because the construction accounting process is so complex, construction professionals who have not completed accounting training should strive for excellence in data collection and timely reporting of all costs and should compare them carefully with the original estimated costs. Any differences are an early indication of potential profits or losses. We prefer to avoid the term *cost control* and instead use the term *cost management* because you cannot change estimated costs after signing a contract, but you can change your reaction to costs that exceed the initial estimated costs. This approach is the essence of cost management.

10.9 Summary

Construction accounting is complicated and is the responsibility of top management, not the independent outside auditors. It is critical that top management understand the accounting process and discipline the accurate and timely input of accounting data from the operations side of the business. This responsibility cannot be abdicated to middle managers or to independent auditors. A qualified and trusted CFO must be in an intricate part of the top management team – listened to and respected.

The objective of most businesses is to grow and be stable entities in their markets. But growth entails risk, which must be managed. One of the largest risks in growing too fast is growing at a rate that outpaces the firm's capacity to sustain the growth. As projects get bigger, margins naturally become smaller, and because contractors essentially handle large amounts of money and get to keep only a very small portion, it is critical to keep a close eye on profit margins to ensure that growth happens in response to profitable operations and contributes to growth in the net worth of the business.

Unfortunately, the accounting practices required in construction companies can be complex, and the practices become more complex as a company grows. Consequently, a contractor and top managers tend to focus on getting the projects completed and delegate the responsibility for financial reporting and financial management to others in the company. As the volume of work increases, so does the size of the payments received for the work, but so does the cost of the work and the size of the infrastructure (people, facilities, and equipments) required to complete the work. As the money flowing through the business increases, profit margins on larger contracts become smaller, and it is easy to overlook the erosion of profitability and the business's overall net worth. Decreases in profit margin are most apparent when the market slows down and there is no longer a sufficient amount of work in progress to hide poorly performing projects.

What is worse is that accounting for work in progress is based on estimates, which can be highly subjective. Nevertheless, accounting for work in progress is the most effective accounting method for construction businesses, in which a significant portion of revenue during a fiscal year is composed of work in progress. For the accounting to be effective, it needs to be widely understood by those in the company who contribute to the input and estimation of work in progress, and the entire accounting process needs to be closely managed by the executive team.

11

Construction Industry Cycles

Generally, entrepreneurs are optimists, and owners of construction firms are no exception. Optimism and the drive for growth that prevails in the construction industry appear to have prevented some construction professionals from recognizing that the construction market is cyclical. The market has grown continually since World War II (WWII), but the growth has been interrupted roughly every 10 years (Figure 11.1). The construction market has cycled into decline eight times since the end of WWII. The downturn lasted one or two years on average, and the market then returned to growth until another downturn began. Most construction professionals are surprised or even shocked to hear about this cycle, and others refuse to believe it. It is curious that this reality is not more widely known and planned for. As this book is being written, the industry is enjoying the longest growth streak since WWII.

Construction professionals may produce the built environment, but they have absolutely no influence over the demand for their work. Market cycles are caused by internal or external forces, including overbuilding, labor shortages, economic depression, and international conflicts. Cycles are inevitable. Once construction professionals understand the cyclical nature of the industry, they need to ask whether they should plan for the phases of a cycle. When the construction market slows down, the competition heats up almost instantly because contractors refuse to give up hard-earned gains in sales volume. This pattern has occurred in every market downturn since WWII, driving profits down and, in many cases, even resulting in losses. The question arises: If these cycles are fairly regular and predictable, why do they cause so many project losses and business failures? The question arises: Is there some way that rational business owners can deal with regular and relatively short market downturns?

It may be enlightening to review the dynamics of previous market cycles. During one cycle, the industry had just passed through a year in which contractors collectively had put a billion dollars of work in place in the United States. The industry was celebrating boom years, and contractors were enjoying significant profits. Growth peaked a year later at $1.1 billion. The following year, the industry slowed to just under $1 billion, and profits plummeted nationwide. If contractors made significant profits building $1 billion in projects two years prior, why do the same contractors suffer when they again built approximately the same amount in projects during a market decline?

This phenomenon can be explained. When a construction company's backlog diminishes, the pressure is on to capture more work in order to at least maintain current sales. The situation feels like a crisis, regardless of the economic circumstances, which are simply a cyclical market downturn. Surviving a

The Business of Construction Contracting: Schleifer's Guide to Financial Success, First Edition. Thomas C. Schleifer and Aaron B. Cohen.
© 2025 John Wiley & Sons, Inc. Published 2025 by John Wiley & Sons, Inc.

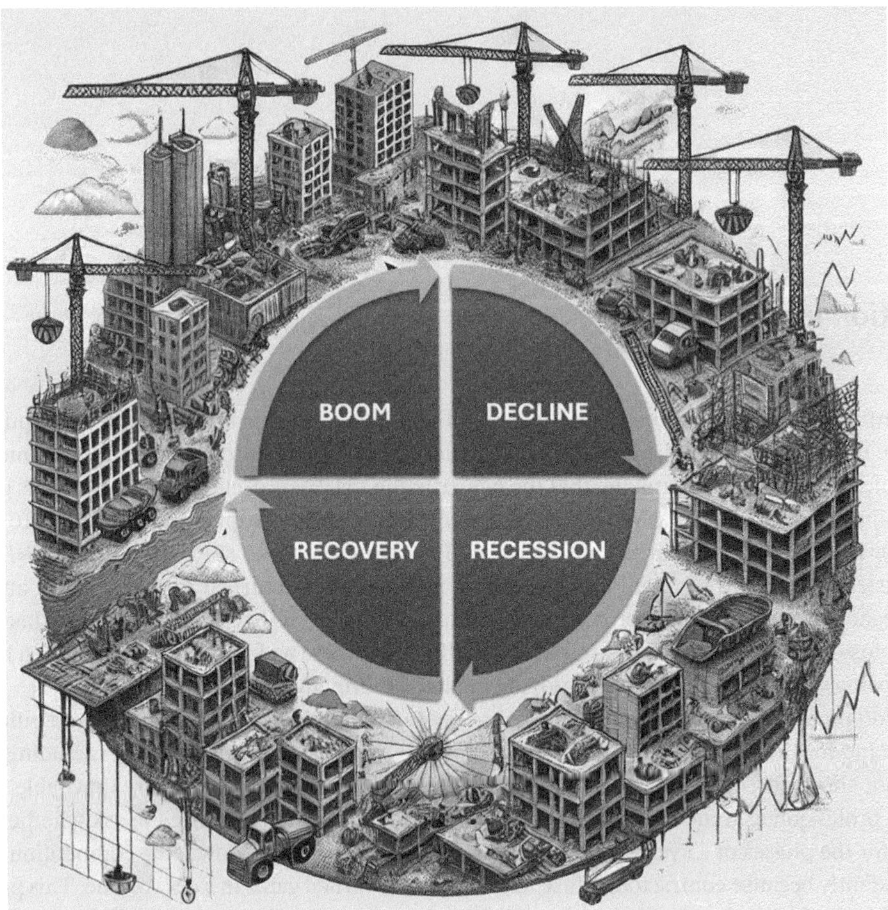

Figure 11.1 The construction market is cyclical, going into a recession roughly every 10 years.

market downturn starts with recognizing that as soon as the market softens, competition intensifies, and prices and potential profits diminish.

What would happen during a market decline if each contractor planned to reduce their sales and capture just their prior share of the current market? Theoretically, with the competitive balance unchanged, each contractor would continue to mark up the work as in the past, and profit margins would be maintained. (*We sometimes refer to this scenario as construction in heaven.*)

11.1 Market Decline

Trying to maintain volume in a declining market is, in effect, attempting to increase market share. An increase in market share is almost always bought at a cost. Contractors who lower their prices to resist even a small reduction in sales usually end up with a lot of cheap work. Consequently, risks increase during an already difficult circumstance. The lower-risk alternative is to accommodate the market and downsize to align the organization with market realities. It makes no business sense to take on cheap work when the potential for risk is the same as for profitable work.

Every contractor is responsible for managing risks during all market circumstances. Construction company owners cannot control the market, but they can control their responses to it by implementing processes they can control. Unfortunately, the appropriate reactions typically range from difficult to distasteful. It is difficult but necessary to keep emotions out of business decisions. Acting responsibly may involve taking steps to temporarily downsize, including rapidly reducing overhead. To effectively control risk, these steps need to be taken sooner rather than later. By far, the largest construction industry's overhead cost is employee wages. It will not be enough to cut back on costs such as subscriptions, bonuses, travel, and entertainment. These costs are minimal compared to wages. Nevertheless, it is appropriate to cut these minimal costs because it signals to all employees what the company's attitude is during a downturn. A reduction in sales means producing less work, which demands a reduction in field personnel and associated field overhead. This reduction, in turn, demands a reduction in management and administrative personnel.

We regularly hear contractors make statements such as the following: "*I can't operate with 10% or 20% less work. I have a drop-dead volume I have to maintain to be viable.*" We respond, "*As you grew your business from $5 million to $10 million and then to $15 million, were you profitable at $5 million and $10 million?*" Most contractors were. The same question applies to $50 *million*, $100 *million*, or any size company. The problem is that to profit at lower volumes, the organization must be reconfigured or resized to what it was when it was profitable at that smaller sales level. Doing so would obviously require serious cuts in overhead and may include selling equipment or mothballing equipment if selling is not practical. Historically, construction professionals have refused this alternative, always hoping the downturn will be short.

Of course, it is impossible to know in advance that a downturn will be short, so contractors should not make decisions based on hope. Our experience indicates that resources retained at great expense are too often eventually let go anyway when a downturn drags on. There is little risk in downsizing because if the downturn is short-lived, the people and equipment eliminated will be readily available to begin working again. The alternative of waiting just months or longer to downsize creates an unnecessary drain on funds that will be critically necessary to finance recovery when the market rebounds. Accurately predicting when a market will rebound is impossible, and trying to wait out the downturn without decreasing overhead violates the primary objective of avoiding loss during a downturn. Profit margins may decrease, but incurring losses makes no business sense at all and must be avoided.

We fully understand that contractors resist laying off good employees because attracting, training, and retaining them is expensive and time-consuming. However, holding on to redundant and unnecessary employees during a downturn is also expensive. Any fear of reacting too quickly should be overcome by thinking of the significant costs of reacting too late. The risk a contractor incurs by not reacting quickly can cause the business to lose everything. Not reacting quickly can be described as business suicide.

It might be helpful to remember that managing risks to the organization protects the jobs of the employees who remain. Downsizing also provides an opportunity to weed out less productive employees. Even when you employ only good people, there are always *good, better, and best.* When forced to lay off during a downturn, it goes without saying that good people will be let go, which leaves the better and best still employed. Clearly, a net improvement. This appropriate response to a market downturn enables contractors to profit on a reduced volume of work while retaining the organization's best people. Managers quickly learn that it is much easier and more satisfying to manage during growth periods than during market cycle downturns.

11.2 Market Recovery

Our study of prior market rebounds indicates that a recovering market can also be a financial struggle for many construction enterprises. One reason is that they experienced serious financial strain during the downturn. Another reason is that aggressive pricing and bidding alwayscontinued until the appetites of the construction organizations that suffered during the downturn were satisfied. Depending on the length and depth of the downturn, this aggressive pricing can go on for years after market recovery begins, delaying margin recoveries for the companies that have been struggling through the downturn.

Research also confirms that the rate of failure among construction enterprises is actually worse during recovery than during the downturn because an increase in sales increases cash flow needs. Each month an organization puts more work in place than it did the previous month, the outflow of cash becomes greater than the incoming cash from the previous month. The result is continuing and worsening negative cash flow. Because cash flow needs increase with increased sales, construction organizations should develop comprehensive strategies to manage cash judiciously during recovery to remain financially viable and creditworthy. Effectively managing cash flow requires the discipline to resist engaging in the frenzy to get more work during early recovery. While it is a natural instinct to try to recover lost volume as quickly as possible, it is not practical because your competitors have the same inclination.

Market dynamics also tend to change during downturns. In addition to the heightened competition and aggressive pricing, clients' preferences during a recovery following a downturn tend to shift from alternative contracting methods back to the low-bid process. This is because project owners become accustomed to the competitive pricing during a downturn, which enabled them to get more done for the money they spent. They then develop capital budgets based on those lower prices. Then when costs for construction services tend to go up during a recovering market, forcing competitive bidding can help keep the cost of their projects closer to the price points they have enjoyed during the downturn. Profit margins following a market cycle seldom return to exactly the same level they were at before the downturn. This ratchet effect has resulted in a steady decline in profits in the industry since WWII, which is part of the reason for today's abysmal profit percentages.

Another issue is that if a company's capital is usually depleted and equity and bonding capacity diminished during a downturn, this causes a firm to be less able to participate in the recovery. Compounding the problem is the fact that over the last 80 years of market cycles, aggressive bidding persisted during each recovery until the total size of the construction market returned to close to their prior size before the downturn. Aggressive bidding compresses margins when there is a need for them to increase the most. In contrast, during the longer growth phases of the market, the increasing amounts of profitable work enable contractors to finance the growth with their increases in earnings.

Absent increased profits during early recovery, growth requires outside financing, which is problematic for firms financially weakened during the downturn. Adding to this issue is the uncertainty that the banking industry usually displays following a downturn. This uncertainty inevitably delays construction lending, which combined with the financial strain following a downturn, becomes the perfect storm. This is the reason that a recovering market regularly causes more construction business failures than the preceding declining market.

11.3 Lessons Learned

The prime objective of being in business is to earn a profit and generate a return on investment for the company's shareholders. To accomplish this during a declining market requires deliberate, strategic downsizing and delaying such actions only magnifies the risk. This is a hard lesson. Reducing a company's workforce is difficult, so much so that repeating what we have already stated on this topic seems appropriate. Surviving during a market downturn and the recovery phase in the construction industry starts with recognizing what is happening in the marketplace. The results are predictable because they have occurred without fail during every industry cycle for the last 80 years. When there are fewer projects in the market, competition intensifies, and prices and profit potential diminish. In a shrinking market, the ideal response is for each contractor to accept less work so that each business maintains its market share. However, contractors tend to resist, often strenuously, any reduction in sales and to fight vigorously for the fewer available projects, intensifying the competition for the work and driving down prices for everyone.

This tendency persists perhaps even more strongly through the initial and intermediate stages of recovery when every project still looks like the last project a firm might see. Some firms try to be the first business to return to its prior size. By doing so, the firm is trying to increase its market share, which limits profit potential and magnifies risk in an already risky environment. The risk is compounded by the fact that in each stage of recovery, particularly after a longer downturn, inflation is common because of temporary labor and material shortages.

Some construction tradespeople and managers who were laid off and left the industry do not return or return only after becoming confident that the recovery will be sustained. Material manufacturers and suppliers have suffered the same financial difficulties as contractors, having been forced to cut back capacity and unable to raise prices while incurring the additional costs of carrying inventory. Regaining capacity for material suppliers requires investment and time, causing material shortages during recoveries that lead to rapid price increases. Contractors cannot control the market, but they can control their response to it. Market conditions during recoveries are disrupted and unpredictable. This, combined with aggressive pricing to regain lost volume, is all the more reason to resist trying to be first to recover.

11.4 Company Downsizing

Construction professionals should fully understand the potential for success or failure in the contracting business during market decline and how to downsize a construction business when necessary. This means being prepared to downsize. In a country in which bigger means better, the idea of downsizing is loaded with negative connotations. However, in a construction market downturn, downsizing is a profitable alternative to fighting for every job, thereby diminishing profits. Downsizing is a deliberate shrinkage of a company's direct costs and overhead in anticipation of a decline in sales. Downsizing is the intelligent response in an economy that contractors have no control over (Figure 11.2). For example, downsizing generates cash because the money comes in from the greater amount of prior work and because current spending is lower because of the decreased amount of current work.

It takes a while to become a good planner and to become proficient at forecasting. Planning and forecasting are important in order to guard against unwarranted optimism and to accept downturns as

Figure 11.2 Downsizing is an intelligent response in a market downturn to protect profit margins as sales volume declines.

natural. Downsizing decisions are never easy, and it is almost impossible to make good decisions on short notice. One of the benefits of a downsizing plan is that decisions are not developed under pressure but as part of a hypothetical exercise. Therefore, decisions that seem impossible to make during a time of distress are seen in a different light.

A contractor that sees a market decline coming and decides it is healthier and safer to operate at a reduced sales volume will immediately reduce overhead to align with the anticipated reduced volume. Because overhead consists mostly of personnel costs, the main task is to determine who will be let go. Most construction owners will initially think that they cannot let anyone go, and that they need every person for the business to function. The reality is that everyone employs good people. Some are better than others, and some may even be essential. A downsizing plan should categorize key employees to determine which are to be laid off and in what order. Downsizing increases efficiency and productivity, because the preplanned employees are laid off in appropriate order. Top managers get closer to the work when layers of managers are reduced, and remaining employees understand that extra effort may be necessary. Most construction professionals work long hours already, so during downsizing, it is important to ensure that work is allocated appropriately to the remaining employees.

The next question is to determine which equipment needs to be sold or mothballed. Mothballing the equipment is not the most efficient option, because the equipment will continue to age and there will be continuing costs of ownership, such as taxes and insurance. As with employees, equipment can be ranked so that when a downturn comes, it will be much easier to determine which equipment to sell and what to do with other equipment.

11.5 Rightsizing

With a comprehensive plan, downsizing a construction firm is not complicated, and a number of contractors have downsized to accomplish what we call *rightsizing*. Rightsizing is adjusting the size of the firm to the size its owner considers appropriate. What is deemed an appropriate size should be driven by market conditions or by internal, usually personal reasons. We have observed many contractors that achieved a profitable and comfortable sales volume and found that when they exceeded that volume, they experienced diminishing returns. Other firms were in markets with limited room for expansion, and the firms elected to manage their size rather than take on the risk of expanding their geographic areas. These contractors rightsized their businesses.

We are familiar with a considerable number of contractors that realized they had exceeded their management capacity. Instead of hiring more managers, these companies rightsized their operations to a manageable, profitable size and remained there indefinitely. In our experience, contractors who choose to maintain a certain profitable size usually prosper. Absent the effort and cost of growth, a construction company that maintains a consistent size tends to get continually better at what the company does. Most of these firms enjoy a higher profit percentage than firms that continue to grow.

A Midwestern contractor rightsized its business at $35 million annual sales and had a reputation of being almost impossible to compete with in its limited market area. Many competitors chose not to participate in bid requests after this contractor was added to a bid list. Consequently, there were often only two or three bidders when this contractor went after a project, and in the majority of instances, the contractor was the successful bidder. The contractor's annual profit margins were some of the highest its surety had ever seen.

A southeastern contractor maintained the same sales volume for three decades and was difficult to beat in its home area, to the point that it had an amazing success rate. We refer to this rate as their hit rate. In our experience, contractors at that time competed, on average, on 6 to 10 projects to capture one project, which translated to a hit rate of 1 in 6 or 1 in 10. The contractor in question had a *hit rate* of 1 in 2.5 projects. In our experience, this kind of success rate can be accomplished only by a company that performs the same amount of work with the same employees over long periods and just continues to become more and more efficient. Not growing allows an organization to do the same thing over and over again, and repetition is an excellent learning and process improvement method.

11.6 Overhead Research

The biggest drawback to coping with market cycles is that our industry has always been driven by growth. Therefore, construction organizations have been forced into the posture, *"If you're not growing, you are going backwards."* To test this, we ask contractor audiences, *"If the type of work you do was for any reason no longer required, like those who used to make buggy whips, what would you do?"* Universally, the answer was: *We would go into another type of construction.* It is clear that maintaining and growing sales is a bedrock principle in the construction industry.

With growth as a prevailing mandate, proposing downsizing or rightsizing as a defensive reaction is a hard strategy for contractors to embrace. Because growth is paramount, industry acceptance of the indisputable data that the construction market is cyclical has limited acceptance. The prevailing belief is that continuous growth is the only viable business plan for a construction organization and that expanding overhead is an appropriate step in that direction. Overhead costs are also commonly considered permanent costs.

Construction industry beliefs need to change starting with the acceptance that the successful business model for all major industries in the United States is profit – not growth. There is nothing wrong with growth, as long as it is profitable growth. In a cyclical market like construction, profitable work is possible during the longer growth periods of the cycles, but becomes less profitable with some losses during the shorter declining market phases. During high-profit years in the past, it was easier to withstand earnings reductions, even modest losses, during the downturn periods. With the steady erosion of profit margins over the last several decades, the reduced earnings no longer cover the losses during market cycles decline. This reality prompted a multi-year research project that resulted in what we believe is the only solution possible – *Flexible Overhead*. It requires a paradigm adjustment in managing overhead that fortunately increases profit and reduces risk.

11.7 Flexible Overhead

For many people, adopting a flexible overhead approach will require a paradigm shift because the current growth model in the construction industry supports increasing fixed overhead in order to continually expand capacity. Flexible overhead addresses the misconception that you cannot get half a person, half a truck, or half a piece of equipment. You can. The key is to have 15–25% of overhead expenses engaged in a way that they can be turned off within a day, a week, or a month. Examples include rented equipment, temporary personnel, and rented office and shop space.

Initial reactions to this approach may be that it is impractical, that temporary employees are less qualified, that it is cheaper to own than to rent, and so on. However, we must have an alternative because construction companies cannot continue to operate the way they currently are, which is decreasing overall profits during market downturns by taking on losing projects, in an attempt to avoid any reduction in sales. All businesses have the objective of earning a profit on all the work put in place. If all overhead is permanent, then a minimum sales volume must be maintained at all times in order to support the overhead costs. Of course, maintaining this volume is impossible in a declining market, so overall profit is almost guaranteed to decline and might even disappear completely. Striving to maintain a certain sales volume is also what drives contractors to take work they have limited experience with or work that might not be a good fit for the organization.

If there is any extra cost from flexible overhead, and there seldom is, it can be considered low-cost insurance against the costs of fixed overhead when the overhead is not needed. An advantage of flexible overhead is that the contractor has complete control of their overhead and the luxury of selecting only projects that the company can complete at a profit. The objective is not to be forced to take on high-risk or losing projects in an attempt to maintain permanent overhead. This is reinforced by the certainty that profits can be maximized by taking on a smaller amount of work. Consider this reality check: Is there one project completed in any prior year that you wish your company had not taken? If the answer is yes, you are in effect saying that you would have preferred to have lower sales that year and more profit.

11.7.1 Overhead Survey

Following the extended market decline in the 1990s, a survey involving almost 500 construction enterprises was conducted. The results verified that most construction enterprises maintained overhead for as long as possible during the years-long decline: 98% of respondents maintained 100% of their overhead for at least the first year of the market decline. By the end of the multi-year decline, 92% of respondents had cut one or more categories of overhead. Of these respondents, 73% reduced bonuses, company functions, and charitable donations, and 41% reduced business development overhead. Further, 24% of respondents cut overhead by as much as 25%, while 15% of respondents cut overhead by more than 50%. In addition to the cuts already mentioned, cuts were made to salaries, training, education, and travel; particularly large cuts were made to retirement plans.

Our evaluation of these data suggest that every category of overhead expense should be scrutinized regularly. Overhead costs need to be adjusted annually (or more often) to align with sales projections for the following year. Further, the data indicate that all of the enterprises should have cut overhead costs sooner: Almost all eventually cut overheads but only after significant reductions in profitability and some after very serious losses. The sacrifices in profitability could have been avoided by cutting overhead sooner. Companies should not wait to reduce overhead until a market downturn occurs. Instead, companies should start reducing overhead as soon as possible, even when a downturn seems likely but has not yet started. The question we are concerned with is: Will the industry absorb the critical lesson that overhead needs to be put in place with greater care, that overhead should be reduced sooner, and that some portions of overhead should be flexible and easy to cut on short notice?

We are convinced that the profitable contractor of the future will manage overhead through much stricter controls and ease the pain of cutting overhead by maintaining 15–25% of all overhead as flexible overhead, which can be turned off on short notice, in many cases, a week or less. A flexible overhead process works well with the construction industry's cyclical market, in which an enterprise may do its regular volume one year, 120% of that volume another year, and 80% another year. Permanent overhead requires a stable market or high enough profits to survive market downturns. The market has never been stable, but from the end of WWII until about four decades ago, profits were high enough for companies to survive market declines. The high profits have evaporated, and the market has remained cyclical. With lower profit margins, contractors must accept market realities and adjust business practices accordingly.

Why have the declining profits over decades not been obvious to most contractors? Because profits have decreased slowly, and most construction professionals believed that the industry was just becoming more difficult. For years, a number of practitioners have described the business as not being as much fun as it used to be. Research on the industry over decades has made the pattern of declining profits clear. Finding the trend distasteful and wishing it was not so will not change the situation.

Overhead is not necessarily negative or positive. It is a cost of doing business, and efficiency requires that costs be minimized to the extent practical. The determinant must always be how the overhead will advance productivity and whether the overhead is for something that is part of a company's core competencies or whether another process can accomplish the same thing better and more economically. When overhead is controlled, a contractor has the luxury to work on only the types of projects it is experienced with. Therefore, the contractor can better manage risk, dramatically reduce the number of losing projects, and maximize profits.

Employee wages and benefits are generally the single largest overhead cost category for a construction organization and are typically considered the most permanent portion of overhead. Many company owners state, "*It took forever to find these people and a long time to train them. We have a lot invested in them and can't afford to lose any just because work is slow.*" We do not argue with this position. However, what many companies do not recognize is that many overhead office positions have become highly specialized, and this specialization is not necessary or efficient (Figure 11.3).

Case Study: Overhead

The best way to explain this situation may be to present a case study of a construction company that was introduced to and embraced flexible overhead quite some years ago. This company had four highly skilled clerks in the accounting department. Each clerk managed a single category of the work: one clerk handled accounts payable; another, accounts receivable; another, payroll; and another, the general ledger. All reported to the CFO and his assistant. The company was in the upper portion of the United States, and during fair weather, the company was very busy. In the summer, the company hired several college students to help with the workload in the accounting department. During the five or six months of colder weather, there was never enough work to keep the four full-time clerks busy, but everyone thought this situation was just part of the business. The founder had organized the business this way, and his successors maintained the status quo when they took over the business.

The authors conducted a brief study and determined that the accounting workload in the winter could be accomplished by two people. We advised the company to cross-train the two most

Figure 11.3 Temporary employees can be used for 15–25% of contractors' overhead staff so they can be more easily let go when work is slow and replaced when the work picks back up.

productive clerks and lay off the other two. When the workload picked up in the spring, the company obtained two accountants through a temporary agency to work part-time. By the summer, the temporary employees worked full-time, and then their hours were reduced again to part-time in the fall. They were not needed during the winter months. One of the major advantages of getting employees from temporary agencies is the employees can start and end on short notice. Additionally, temporary workers' benefits are paid by the agency, so their actual cost is competitive. Further, the company's costs are lower in the long run because the temporary employees are engaged only when needed. The case study company cut its total accounting overhead costs by more than half, and efficiency went up considerably. Everyone concerned hated the idea when it was put in place, but they loved it a year later.

We then looked at the estimating department. Top managers said that the six estimators were always busy and efficiency was at a maximum. However, a two-week study determined that the estimators spent a little less than half of their time doing actual estimating work and attending occasional meetings. They spent the remainder of their time making copies, obtaining material details by looking in catalogs or making telephone calls, hunting for or sorting through plans specifications and other documents, and completing other clerical tasks. When these activities were discussed after the study concluded, all the estimators said they wished someone else could do the non-estimating work.

Improvements were quite simple. The company reduced the number of estimators to four and hired two clerks to complete the non-estimating tasks for the estimators. The company saved a substantial amount of money, and efficiency went way up. Perhaps even more significantly, job satisfaction improved immensely. Some may ask: *"what about the two estimators that were no longer needed?"* Whether they were reassigned or laid off is not the focus here. If managers are not willing to improve efficiency and profitability by examining overhead costs, they need to revisit their priorities. Hopefully, top management fully understands that inefficiency affects their competitive position.

11.8 Expanding Flexibility

This same process can be used in every department to improve efficiency and fine-tune overhead costs. In many cases, it will be appropriate to outsource work that a highly specialized firm can provide at a lower cost than can be achieved in-house. A large number of functions can be outsourced to reduce costs and increase efficiency. For example, outsourcing the preparation of payroll may be the easiest overhead savings possible. Oddly enough, construction companies seem to strongly resist outsourcing this function. Some owners argue that they work with multiple unions; the benefits and wages differ from job to job; in some places, union dues have to be deducted; and so on. Payroll companies are experts in these areas, and in our experience, 100% of companies that have turned payroll over to a third party were perfectly satisfied with the transition and reaped immense savings. In most cases, expenses for payroll preparation were decreased by more than 50%, and this percentage does not take into account the cost savings of not having to provide office and parking space for the employees that used to handle payroll.

We know flexible overhead works because hundreds, if not thousands, of construction enterprises are using it. Explaining the "how to" to readers is not nearly as important as explaining the "why." The simple explanation is that permanent overhead in a cyclical and modest profit industry like construction just does not work. Permanent overhead worked in the past because profit margins were much higher and the necessary rental, temporary employee, and services-for-hire infrastructure were not readily available. All that has changed, while the failure rate in the construction industry has increased.

We could go on and on about how to implement flexible overhead, but we know that our readers can figure that out easily, so we will stay with the why switch to flexible overhead. In a word: "SURVIVAL." The risks in our industry are greater simply because the rewards have been structurally reduced. The primary source of profit is putting the work in place. In the past, you needed to own the equipment and facilities to accomplish that, which increased your costs, risk, and exposure. Today, the skill set you bring to the business is accomplishing the work. Concentrating on that is easier when you don't spend much of your effort buying, owning, transporting, and preparing tools and equipment. There is little profit to be earned in producing payroll, completing every office task, and so many other tasks that can be outsourced, causing a greater percentage of your productive employee's time focused on earning a profit. A simple analogy is that you can send someone to a vendor to get what you need or simply have it delivered. That is what flexible overhead is, and you only pay for what you need, which means if there is no profitable work available you, just reduce costs.

In our experience, companies generally resist outsourcing because they misunderstand it or fear they will lose some control. In reality, outsourcing works, costs less, and is flexible in that a company pays for only what it uses. Successful contractors of the future will improve how they put construction in place, streamline their overhead costs, and outsource functions that a third party can do better and for a lower price. Companies that do not take these steps will be weaker competitors and will be at greater risk, particularly during market downturns.

11.9 Equipment Ownership

Until the 1970s and 1980s, heavy equipment, such as earth movers and cranes, and smaller equipment, such as chain saws and compressors, had to be owned because of limited rental equipment infrastructure and availability, particularly in suburbs. Fast forward to today, and rental equipment of any type is readily available. Further, leasing or renting heavy equipment may cost no more than owning the equipment and is certainly as efficient (Figure 11.4). Many firms have switched from owning an entire fleet of equipment to renting equipment when needed, and these firms say that one of the things they like about renting is that if a piece of equipment breaks down, the firm simply calls the rental company, and the item is replaced within days if not hours. These contractors recognize that their core competency is producing the work and not maintaining the equipment.

Among the construction enterprises that continue to own equipment, many have altered their fleets as a result of research we published decades ago demonstrating that it is inefficient to own equipment that is not operated at least 60–70% of the year (the percentage varies depending on the type of equipment). Renting, not leasing, the lesser-used equipment is more economical, and this situation is magnified when equipment that is inappropriate for the work is used simply because it is owned. Renting the right equipment is always more productive and often safer than using equipment that is too big or too small for the task at hand just because it is owned.

11.10 Flexibility in Project Selection

When there is not enough work of the size and type that the organization thrives on in the geographic area the organization normally operates in, the organization has a choice: go after projects that the organization has limited or no experience with or reduce overhead, thereby reducing the amount of work required to support the overhead. Calculating in advance which option has the greatest risk is simple. Doing less work and doing only the work the company normally does and profits from has an excellent likelihood of being profitable. The best that can be said about taking on unfamiliar work is that it may make a profit, break even, or lose money. It is actually an unknown. Risks of this nature are best left to gamblers, not businesspeople. If jobs that a company has profitably produced a dozen times can lose money, and we know they can, what are the odds for work the company has not proven successful at?

Figure 11.4 Renting equipment can be as effective as owning equipment and reduces a company's fixed overhead.

Flexible overhead provides options. Absent a desperate need for sales, a construction enterprise is able to go after only good, risk-balanced work because sales will no longer be an organization's only motivation. With flexible overhead, a firm is able to easily reduce or increase overhead practically overnight.

If you wish you had not taken on a specific project, you may want to consider converting some portion of your overhead costs into flexible costs. The other choice is to continue with what is standard in the industry, which is to go after whatever size or type of project, in any geographic area, at whatever low price it takes to capture the project and accept the risk of low profit or a loss to keep the entire organization as it currently is. If a market decline is brief, the company may survive.

The other choice is to reshape the organization's overhead costs so that there are two components: (1) a permanent core group that is always protected, and (2) a nonpermanent group that consists of 15–25% of the total overhead amount and can be discontinued within a week. This flexible overhead facilitates a decrease in sales as well as an increase in sales. Using temporary skilled personnel to expand the capacity of the core group works as well in a declining market as it does in an expanding market.

11.11 False Beliefs

Most of the struggling construction enterprises we are familiar with embrace a business model driven by growth, which puts too much pressure on increasing sales or maintaining volume for the sole purpose of covering overhead. This business model does not work and is totally inappropriate in a cyclical industry. The following common beliefs, which are part of the prevailing model, need to be reconsidered:

- Growth is always good.
- Overhead is a symbol of success.
- Overhead is not to be surrendered unless a company is absolutely forced to.
- Cutting overhead is an admission of failure.
- Downtimes are bad news but a natural part of the industry.
- The industry is not necessarily cyclical.
- Unprofitable work is just part of the business.

The danger of these beliefs is that they are embedded so deeply in the industry that construction professionals do not think to challenge the beliefs, and they drive standard processes. Consequently, companies go after whatever projects are available, in good times or bad.

Another false belief is "*We can do anything.*" Taking on work that a company has limited experience with can result in what we call the *80/20 problem*. In a study of hundreds of failed construction companies, we determined that all of the financially distressed firms had an abundance of profitable work. Their problem was that they also had unprofitable work. Of course, many, if not most, construction enterprises have occasional losing projects, and these projects will not necessarily put a firm out of business. But there is a breaking point. In the companies that failed, an average of 80% of the work was profitable and 20% was unprofitable. The 20% of unprofitable work incurred losses that the profitable work could not compensate for because overhead costs were permanent.

Case Study: Profitable and Unprofitable Work

To further clarify these issues, consider the following case study: A contractor with $10 million in annual sales had $1 million in overhead costs. Eighty percent of the work, or $8 million, generated a 13% profit, totaling $1,040,000. Of this amount, $800,000 was allocated to pay for overhead costs. Therefore, the remaining profit was $240,000. The remaining 20% of the work, or $2 million, generated a loss of 5%, or ($100,000). This loss had to be covered by part of the $240,000 in net profit earned by the 80% of (profitable) work. Thus, the profit was reduced to $140,000. Because the $2 million in unprofitable work did not contribute to paying the overhead of $1 million, the remaining overhead costs had to be paid with the profit from 80% of work. The overhead balance was $200,000, but only $140,000 was remaining from the profit on 80% of work, so the company experienced a loss of ($60,000) over the year (Table 11.1).

Table 11.1 Comparison of a company's financial performance when overhead is fixed and sales volume both includes and excludes $2 million of unprofitable work.

	Profitable work		Unprofitable work		Annual totals	
Sales	$8,000,000		$2,000,000		$10,000,000	
Costs of construction	$6,960,000		$2,100,000		$9,060,000	
Gross profit	$1,040,000	13.0%	−$100,000	−5.0%	$940,000	9.4%
Overhead expenses	$800,000		$200,000		$1,000,000	
Net profit	$240,000		−$300,000		−$60,000	−0.60%

	Profitable work		Unprofitable work		Annual totals	
Sales	$8,000,000		$0		$8,000,000	
Costs of construction	$6,960,000		$0		$6,960,000	
Gross profit	$1,040,000	13.0%	$0	0.0%	$1,040,000	13.0%
Overhead expenses	$1,000,000		$0		$1,000,000	
Net profit	$40,000		$0		$40,000	0.50%

If taking on unprofitable work is part of being in the construction business (and we do not agree that it is), then minimizing the amount of unprofitable work must be a top priority for a company's senior managers. (*This company earned a lot of money and then wasted it on losses.*)

11.12 Losing Projects

We spent years researching, defining, and measuring the causes of construction business failures. This work led to research on how to manage losing projects while they are in progress, with the goal of reversing the losses. Our research confirmed that it is:

- incredibly difficult to turn losing projects into profitable ones
- it is hugely expensive to manage losing projects
- it is difficult to eliminate the damage done
- it is impossible to minimize costs already incurred.

Once loss is discovered, it may be possible to slow the rate of loss through massive, concentrated effort. But it is almost impossible for the remainder of the project to earn enough to cover losses already incurred, because the remainder of the work would have to somehow become exceptionally profitable.

We could not find any examples of companies that accomplished this. The only effective strategy is to avoid taking on unprofitable projects. Not taking on potentially unprofitable projects requires huge effort to recognize what might be a losing project, but the effort could prevent a company from failing.

Through our extensive research, we also discovered that there are no bad projects, just bad matches between contractors and projects. This finding is significant because the construction professionals with losing projects generally blamed the problem on the project, the designers, the weather, the supply chain, and so on. In almost all cases, we could find similar projects that were successful. It became apparent that the low-bid system and a mismatch between the contractor and the project were the primary causes of projects losing money. The third- or fourth-highest bidder may have been a better match for a project than the contractor that was awarded the project. Too often, the low bidder is low because they do not understand the project or the risks or both.

Losing projects do not just happen. They are jobs that are deliberately pursued and captured, typically by an organization with limited or no experience that qualifies the organization to estimate, execute, and finance the work successfully. In the construction industry, gaining experience is a paradox of industrial proportions, because if experience is critical to project success, how does a firm gain the experience needed to complete a project profitably? Along these lines, should a construction organization ever take on a type of work the organization has never performed before? This issue is a matter of risk tolerance and decisions about prioritizing profit or volume. Which to prioritize is clearly a contractor's choice to make, but decision-makers should understand that they must buy and pay for experience.

Experience and knowledge can be extremely expensive, considering the serious risks associated with inexperience. When a company with limited experience in a type of work attempts the work, the company should expect the unexpected. There are few effective strategies other than to start small, finish the first project before attempting another, and attempt only what the firm can afford to lose. It is extremely difficult to measure the risk of inexperience, but suffice to say, the risk is huge.

11.13 The Importance of Experience

There are three critical elements involved in achieving project success: experience, experience, and experience. Project risk is directly correlated with an organization's prior experience with similar work. Therefore, project risk is different for every company. An important discovery in our research was that project risk can be measured. Unlike in manufacturing, construction projects are rarely duplicated. Most projects differ in significant ways. Limited replication confines experience to similar projects that have previously been priced, built, and completed at a profit. Direct experience with the size, location, type, and design of the work are critical to success. Complicating this situation is the reality that, very seldom, are the employee teams exactly the same from one project to the next.

The more similar a new project is to previous successful projects, the more likely the estimated performance will be achieved. The reason is that planning and execution are more of the same as opposed to totally unique. Experience is accumulated institutionally but captured individually. Consequently, the number of team members with direct experience on similar projects is hugely important and can dramatically affect the likelihood of achieving estimated or improved project performance.

11.14 Summary

The construction industry is cyclical, and those cycles tend to follow a repeating pattern that is fairly predictable. However, predicting when precisely the market may take a downturn is not the point of this chapter, rather the intent is to communicate the importance of being prepared to respond appropriately to the downturn when it eventually does happen. Most contractors are growth-oriented, and while that may even be core to an organization's mission, a construction business is not a business if it fails to make a profit. Responding to a declining market by attempting to take on more work or different types of work in different areas, and in intensified competitive conditions for the sake of maintaining a desired level of sales volume can only result in unprofitable work.

Striving to maintain prerecession levels during a recession greatly increases the risk to the company's ability to weather the storm and be prepared for when the market does rebound. Unfortunately, when contractors try to maintain their size at all costs, they can become so financially strained during a downturn that when the market does rebound, they lack the financial strength to compete and risk failure. Contractors need to recognize a downturn when it is occurring and adjust to what the market will bear. In a declining market, a company should adopt a strategy of decreasing sales volume and adjusting company overhead to the level of volume that is sustainable for the current market conditions.

12

The Science of Project Selection

It is not uncommon to hear people in the construction industry say, *"If you're not growing, you're going backward."* Most of the construction enterprises we are familiar with embrace (or are strongly influenced by) a business model driven by growth. However, the business growth model that most people in the industry are familiar with is totally inappropriate in a cyclical industry. This model puts much too much emphasis on increasing or maintaining sales volume for the purpose of covering overhead, no matter what the market conditions are.

Construction professionals seem to have limited interest in market share fundamentals, which include propositions such as the following: *If the market you are working in is growing, for example, at 5%, each business in the market can expect to grow at ±5% while maintaining its normal markup and profit.* Note the illustration in Figure 12.1.

If a business in the market grows at a rate greater than 5%, then other contractors in the market will have to grow less than 5%, because there is only so much work available, as illustrated in Figure 12.2.

In other words, the only way to increase market share beyond the market rate of growth *is to buy and pay for it* by taking work from competitors. This concept is easier to understand in a product sales context. Take, for example, the cosmetic industry. If a cosmetics firm believes it can expand its market share from 10 to 15% in a particular metropolitan area, the firm will typically put together an aggressive advertising campaign, which usually includes reducing the prices of products.

The promotion will be planned to last for a specific length of time or until certain goals are achieved. If the goal is to increase market share to 15%, the campaign will last until the firm has achieved some predetermined percentage greater than 15%, perhaps 17 or 18%. The reason for this is that marketers and firms have learned through experience that some small percentage of buyers who are attracted by the sales will not continue to purchase the firm's products when the prices return to normal. The advertising and price reductions cost money, and the firm may operate at a loss for the length of a campaign. If management's efforts are successful, they will have bought and paid for an increased market share, and the costs will be more than recouped through the increased volume of recurring revenue gained through the market expansion.

The lesson should be clear. Increases in market share cost money. Increasing market share is not free or easy. The problem in construction is that the results of lowering prices to increase market share do not work because the results are temporary. There is considerably less customer loyalty to construction

The Business of Construction Contracting: Schleifer's Guide to Financial Success, First Edition. Thomas C. Schleifer and Aaron B. Cohen.
© 2025 John Wiley & Sons, Inc. Published 2025 by John Wiley & Sons, Inc.

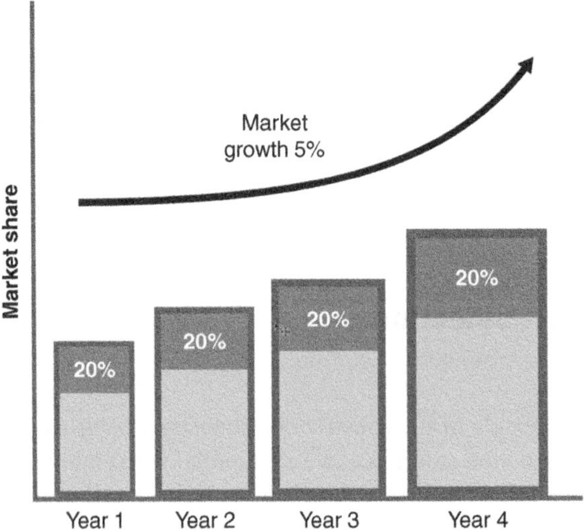

Figure 12.1 Contractors' share of the market keeping pace with 5% market growth.

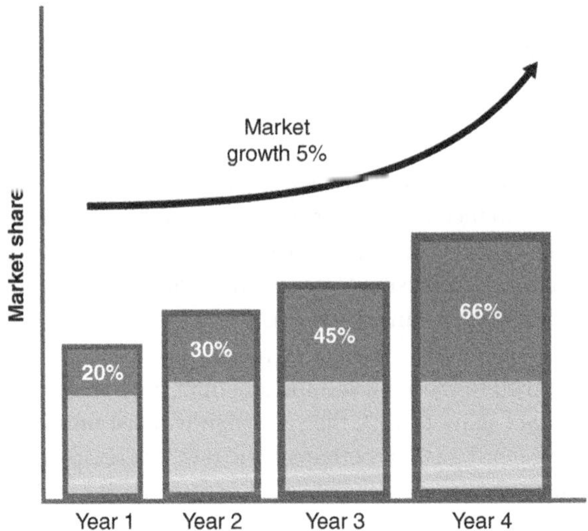

Figure 12.2 Contractor capturing a greater share of the market from competitors by growing 10% each year, while the overall market growth is only 5%.

companies than there is to the brands of retail products and the companies that sell them. Price continues to drive buyers of construction services, despite the development of numerous contracting methods that are alternatives to the low-price bid approach. In addition, growth in the construction business may increase the amount of profit but often decreases the percentage of profit. Any decrease in profit percentage increases business risk. Earnings data going back generations verifies that small contractors earn a greater percentage of profit than midsize contractors, which generate a greater percentage of profit than large contractors. We are not antigrowth. Our intention is to explain the growth dynamics prevalent in the construction industry.

When the authors entered the construction industry some years ago, there existed a belief that some projects would be losers no matter which company ended up constructing them. For generations, unprofitable projects were considered random, caused by anomalies within the projects themselves, and unavoidable. An experienced construction professional claimed that *the ratio of bad projects to good projects was about 1 in 10 and advised care because there is a loser out there with your name on it.* This belief seemed to remain unchallenged until we conducted research in the 1980s that showed beyond doubt that there are no bad projects. There are just bad matches between contractors and projects. All of the failed construction firms we examined had completed many profitable projects. These firms were convinced that they were qualified to select, capture, and complete work profitably. However, when the company started to run out of money, the belief proved to be in error because they had also selected projects that lost money.

These findings are now decades old, so it is fair to ask why construction companies have not changed their perspective. This perspective has contributed to the industry's long-standing status as having the second-highest failure rate in the nation. The reason that construction professionals have not accepted our research-backed finding is that the construction industry has been around since the Stone Age, and industry beliefs are difficult to change.

12.1 Industry Beliefs

The belief that some jobs are just losers and can't be avoided is not the only well-established belief in the construction industry that needs to be reconsidered. The following are examples of beliefs prevalent in the construction industry that need to change:

- Growth is always good.
- Overhead is a symbol of success.
- Overhead is not to be surrendered unless a company is absolutely forced to.
- Cutting overhead is an admission of failure.
- Downtimes are bad news, but a natural part of the industry.
- The industry is not necessarily cyclical.
- Unprofitable work is just part of the business.

These beliefs cause many companies to go after any project that is available, in good times or bad, whether or not the organization has experience with the work, the owner, the designer, the size of the job, or the geographic area. A lack of experience with any of these elements dramatically increases risk, yet most contractors refuse to give up the goal of growth.

Taking on work that a company has limited experience with can result in the 80/20 problem. All of the financially stressed construction companies that we examined had an abundance of profitable work, but they had enough unprofitable work to result in overall losses. Many construction enterprises have occasional losing projects. However, we do not believe that losing projects are inevitable, and we know of many successful contractors that have learned to screen out losing jobs during the project-selection process. Our research indicates that most construction organizations are able to survive if their losing projects are no more than 20% of sales. Most that suffer losses of more than 20% will fail. Of course, 20% is not a strict guideline but just a data point, because many construction organizations with losses under 20% have also failed. Therefore, minimizing the amount of unprofitable work must be a priority for senior management. It is important to note that in the failing companies studied, profit reduction was not always the only issue. Excessive overhead expenses were often a major issue.

12.2 Profitable and Unprofitable Work

We spent years studying, defining, and measuring the causes of construction business failures. This work led to research on how to manage losing projects while they are in progress. Our research confirmed that it is:

- incredibly difficult to turn troubled projects into profitable projects
- hugely expensive to manage losing projects
- difficult to eliminate the damage done
- impossible to minimize costs already incurred.

We also verified that another primary cause of project failure is inexperience with the type of work being undertaken. Further, gaining the necessary experience is a problem, not the solution. The only solution we found is prevention – that is, not taking on the losing project.

It bears repeating that there are no bad projects – just bad matches between contractors and projects (Figure 12.3). Losing projects do not just happen. They are jobs that are deliberately pursued and captured, typically by an organization with limited or no experience that qualifies the organization to estimate, execute, and finance the work successfully. Experience is a paradox of industrial proportions, because if experience is critical to success, how does a firm gain the experience needed to complete a project profitably? This question leads to another question: Should a construction enterprise ever take on work it has never performed before? This issue is a matter of risk tolerance and requires making a decision about whether to prioritize profit or volume. Which to prioritize is clearly the contractor's choice to make. Decision-makers should understand that they have to buy and pay for experience and that doing so can be extremely expensive. There are also serious risks associated with the learning process, so it is prudent to expect the unexpected. When a firm attempts work that it has limited experience with, some guidelines may help: start small, finish the first project completely before attempting another, and attempt only what the firm can afford to lose.

The 1,000-plus failed construction firms that we studied earned a huge amount of profit. On average, 80% of their projects were profitable, so prevention appears to include *doing less work*. Obviously, many of the companies in our database would still be in business if they had built only the 80% of projects they undertook that were profitable. Therefore, screening out losing projects, which is what enlightened project selection accomplishes, is not only profitable but a matter of survival. This approach is the ultimate method of risk management.

To make our point clear, consider the following question: Is there one project from any prior year that you wish you had not taken? If the answer is yes, you are, in effect, saying that you would prefer to have had lower sales that year. This reality makes it harder to argue against bringing more science into a construction enterprise.

Figure 12.3 There are no bad projects, just bad matches between contractors and projects.

12.3 Profitable Project Selection

The risk associated with project selection can be accurately measured in advance and is directly associated with a construction organization's experience with similar work. Therefore, project risk differs for each contractor. The performance of an organization improves with repetition. However, a construction company's projects differ in varying ways, so there is not enough repetition to experience much improvement. The closer a new project is to the average of previous projects, the more likely the estimated performance will be achieved, because the planning and execution will be more like the previous projects. The measurement of risk also depends on the experience of the team members involved. Experience is accumulated institutionally but is captured individually, so the number of members on a team with direct experience on similar projects affects the ability to improve performance.

The elements that affect expected performance (and impact risk) include project size, project type, project location, the performance team, the client, unusual features, safety considerations, and available room to work. These elements can be measured and weighted to define the risk and measure anticipated project performance. The more experience with similar projects, the greater the likelihood of a successful estimate, efficient production, and timely completion of the work for a profit. Experience with fewer similar projects translates into a lower level of confidence, recognized or not, and greater risk.

To provide an extreme example, if an organization has been successfully constructing relatively straight-forward office buildings and schools and then attempts its first complex hospital project, the organization would bring no institutional experience to the project and would bring limited, if any, individual experience. The likelihood of successfully pricing and producing the work would be limited, resulting in much higher risk compared to an organization that regularly builds hospitals. The risk can be measured and will obviously be different for each construction organization considering the work. The project, in and of itself, does not define the risk. The risk correlates with the contractor's experience.

An organization needs experience with a type of work and a geographic area in order to understand both the work and the area. Unusual project features, such as curved walls, windows, or roofs, outside of the experience of most organizations, are obvious risk triggers. Unusual or unique projects can and will be built, but they have a steep and costly learning curve, amplifying risk. Regarding geographic area, different locations may have unique requirements concerning labor issues, labor skill levels, subcontractor availability, pricing, owner expectations, and other local customs that may affect how the work is managed or performed. A contractor is unlikely to learn all these differences without the experience of having worked in the geographic area. As mentioned earlier, experience is accumulated institutionally but is captured individually, so the number of team members with experience on similar previous projects affects the likelihood of achieving estimated or improved performance.

A company's institutional experience is not automatically imparted to employees who did not participate in obtaining the experience. Therefore, if an organization has experience in all of the aspects discussed (size, type, location, etc.) but none of the project team members have direct experience, the project risk is high. There is some advantage in having at least some relevant experience within the organization. If things go badly, others in the organization who have relevant experience may be able to assist if they are brought in soon enough. Another aspect of project team experience to consider is whether team members have previously worked together. If they have worked together, the risk is reduced.

These criteria that affect performance (project size, type, location, etc.) can be measured and weighted to estimate the risk the project presents. The criteria can be remeasured and weighted as a project progresses to continue to assess project performance. The following subsections discuss some of the criteria.

12.3.1 Project Size

Construction organizations produce projects of varying sizes. Size, of course, is relative. A large project for one firm may be a small project for another firm. *Small* and *large* are used here in relation to a company's average size of projects. Our experience indicates that companies typically complete fewer small projects than average and large projects, though small projects typically earn the highest profit percentages. Average-size projects are generally profitable but at a lower percentage than small projects. Large projects are usually few in number and earn a smaller profit percentage than average-size projects do.

Small projects are often performed as a service for good clients. Though these projects typically yield the highest profit percentage, there are not enough of them to support the organization. Some contractors refer to these projects as nuisance work, but most contractors admit that small jobs *help pay the rent*. Average-size projects are often the main source of profit and cash flow and are therefore the projects a company relies on and survives on. Average-size projects can be described as the company's *undemanding*, even easy projects, in that most estimators in the organization can price these projects, most superintendents and project managers can build these projects, and these projects almost always perform as expected. Therefore, a company has great confidence with average-size projects, and the risk is low.

Large projects can help a company meet critical mass. These projects support the desire for growth and appeal to the interests and ambitions of key employees. Large projects differ from average-size projects in that not all estimators can price large projects accurately, top managers take an interest in preparing the estimate, and there may be long hours or days of work before the bid is submitted. The greater concern about pricing a large project is indicative of the greater risk. What may not be as clearly understood is that greater attention is also a sign of concern because of limited experience with large projects. Even if the company has successfully completed projects of the same size in the past, the company has not completed as many of these projects as it has completed average-size projects. Therefore, there is less cause for a high level of confidence. The amount of experience with any project correlates with the level of confidence in pricing and expectation of success.

It should also be noted that project size affects the risk on a monetary scale. Small projects that do not do well or lose money are not large enough dollar amounts to have a financial impact on the company. An average-size project that loses will hurt the company but generally will not cause the company to fail, because a number of other average-size projects usually take up the slack. In contrast, a large project that underperforms or loses money has the potential to materially affect the company's financial condition.

12.3.2 Project Type

As with project size, prior experience with the project type correlates with the likelihood of successfully pricing the project and producing it on time and on budget. The likelihood of successfully pricing and producing a new type of project is low, resulting in high risk. If most or all of a firm's prior projects are the same type of work as the new project being considered, the type of work will not increase risk.

12.3.3 Geographic Area

Because construction work is produced differently in different geographic areas, experience working in an area affects the likelihood of success and thus affects the amount of risk to the contractor. An extreme example is a contractor located in the Arizona desert attempting a project in Minnesota during the winter. A less extreme example is a contractor with exclusive experience in rural and suburban areas taking its first project in a large urban area. The project working conditions may be outside the organization's experience and would therefore present a learning curve. The likelihood of success would be lowered, presenting a risk to profitability.

12.3.4   Project Team

As previously explained, if project team members do not have direct experience working on a similar project, then the risk will be higher than if one or more team members have prior experience with the project size, type, location, and so forth.

12.3.5   Unusual Project Features

Most buildings in the United States are rectangular. Most roads and highways run relatively straight for much of their length, and bridges and tunnels are straight to the extent possible. Therefore, the lion's share of experience collected by construction enterprises is with straight lines.

If a potential project has a curved wall, windows, or roof or other unique elements, this project is outside the experience of most organizations. If an organization does have such experience, the odds are that the experience is limited. The same goes for out-of-the-ordinary roads, highways, bridges, and other types of projects. Because of a lack of experience with these types of projects, there is considerable risk in proposing and producing them (Figure 12.4).

One might say that a totally unique project is risky in and of itself and that each firm that attempted it would be equally at risk. This assumption is not accurate. The contractor that has the closest experience to the unique project would have risk, but the risk would be smaller than for a firm with less close experience. If no contractor had any or even somewhat close experience, then the risk would depend solely on the project size as measured in dollars. The contractor for which the project was the smallest relative to the contractor's average-size projects would be at least risk. This is because if the project failed and generated a loss, the loss would have a smaller effect on the contractor than the loss would on a contractor whose average-size project was smaller than the unique project.

We need to emphasize that none of these projects have intrinsic risk. Project risk is exclusively a result of an organization's experience with similar projects. Because an organization's experience resides in its employees, selecting the right personnel for a project is critical. To reiterate what we've previously stated, project risk can be measured in advance. We developed a tool to measure the risk. The *Project Selection Program* accurately measures project risk to a specific enterprise. This tool is discussed in greater detail later in the next chapter.

12.4   Measurement of Project Risk

Three important elements are required to determine project risk: *experience, experience, and experience.* Project risk is directly correlated with an organization's prior experience with similar work. Therefore, the risk is different for every contractor. An important discovery in our research was that project risk can be measured. Unlike in manufacturing, construction projects are rarely duplicative because most projects differ in significant ways. Limited replication confines experience to similar projects that have previously been priced, built, and completed at a profit. Therefore, direct experience with the project size, type, location, and so forth is critical to success.

Figure 12.4 Unusual project features, such as curved walls, can greatly increase the complexity and risk of a project for contractors lacking experience with constructing such projects.

12.5 Impacts of Not Taking on Losing Projects

Obviously, if a company screens out losing projects, sales may fluctuate. However, profit and efficiency will go up, while business risks go down. Fluctuating sales make overhead management critical. The successful contractor of the future will be profitable in good markets and bad markets by adding some science to the management of overheads.

Continuous success in construction requires that the drive for size and growth be replaced with a drive for prosperity, as measured by profitability. Prosperity can be achieved only if a company does not take on losing jobs. To avoid taking on losing jobs requires prudent project selection. In turn, prudent project selection requires judicious management of overhead. For many people, effectively managing overhead will require a paradigm shift because the current growth model suggests that increasing overhead means expanding capacity (overhead). Prudent overhead management and project selection do not stunt growth; they simply switch the focus from increasing sales to increasing profitability. To accomplish this shift in focus, it will be helpful to review the concept of flexible overhead management.

12.6 Flexible Overhead

There is a misconception in the construction industry that a company cannot employ half a person, half a truck, or half of a piece of equipment. A company indeed can. The concept of flexible overhead is to have 15–25% of all overhead expenses engaged in a way that they can be turned off within a month and sometimes within a week, even within days. This portion of general and administrative expenses can be turned off because they are brought on as flexible or temporary expenses. Examples include rented equipment, temporary personnel, rented office and/or shop space, and so on.

This concept, which some would call radical, has been well received, and numerous construction organizations of every size across the country have embraced the concept and reported dramatic profit improvements. Perhaps more significantly, owners of construction companies have said that, for the first time in their careers, they feel in complete control. Other contractors have said they are far more confident about the future because the ever-present threat of a declining market is no longer a challenge for them. Some say they feel relaxed for the first time in years.

Of course, some contractors were skeptical or took longer to test the approach because of concerns that it might be impractical. One feared that temporary employees would be less qualified. Another struggled to accept the fact that, in many cases, it is cheaper to rent than to own. Some thought their cost would increase, but any extra costs arising from flexible overhead, and often there are none, can be considered low-cost insurance against times when overhead needs to be reduced. A primary advantage of flexible overhead is that when a firm has control of overhead, the firm has the luxury to select only projects the firm is experienced with. Consequently, the firm can dramatically reduce the number of losing projects and increase overall company-wide profits.

12.7 Conclusion

There are no bad projects, just bad matches between contractors and the projects they are choosing to pursue. Projects are not inherently risky; rather, the risk a project represents to a company is predicated on the experience a company has with projects of a similar type, size, and location. Project experience is accumulated institutionally but captured individually, so the application of a company's collective experience through its people will determine how effectively a company can price the project and profitably build the work.

While avoiding taking on projects an organization has little experience with is the best way to mitigate the risk, it may not be possible for a company to never attempt a new category of work. Therefore, it should be approached carefully. Start small with one project so the firm can absorb the loss should it occur.

13

The Project Selection Program

Many construction professionals feel strongly that they can build or design anything. We happen to agree. However, the pertinent question is, *"Can you build or design it at a profit?"* Some have concluded that construction is not difficult, but construction at a profit is. There is some truth in that because so many construction projects are pursued and captured, but fail to generate a profit. In a number of surveys, we asked building contractors if they would build a road if a customer wanted them to. Most of the contractors responded yes, and a few stated they certainly would. These responses are why we were not surprised that when we asked the owner of a failing road construction company why he had attempted a bridge project, he said he wanted to expand into that type of work. Likewise, when we asked a residential contractor why he took the commercial project that put him out of business, he said, "I wanted to get some experience in commercial work."

Our research clearly shows that projects within a firm's experience tend to be successful, while projects even modestly outside a firm's experience have serious risks associated with them. This finding leads to the question, *"How do I get the initial experience?"* There are undoubtedly more detailed answers, but our favorite is, *"very carefully."* The potential for risk in project selection is high. There are a number of other things that also affect risk, but they are insignificant compared to experience. Prior experience should be the driving factor in decisions about which projects to pursue. The guiding principle should be to manage risk by taking only risks that the company can afford. Determining what risks the company can afford requires soliciting internal and external input. After that, the company should tread cautiously.

There is much to be said about managing losses, but our research indicates that it is much easier to prevent losses than to manage them. In our study, the primary reason that thousands of contractors failed was that they took on work they never should have undertaken, because they had little or no successful prior experience with that type of work. These projects were failures and caused the companies themselves to fail. Many of these projects were far removed from the types of work the contractors had experience with, but some of the projects were just slightly different from the company's normal projects. It is unfortunate that these firms are no longer in business, but analyzing their struggles enabled us to identify the root cause and develop a solution: the *Project Selection Program* (Figure 13.1).

The Business of Construction Contracting: Schleifer's Guide to Financial Success, First Edition. Thomas C. Schleifer and Aaron B. Cohen.
© 2025 John Wiley & Sons, Inc. Published 2025 by John Wiley & Sons, Inc.

Project Selection Program©2024

v1. © Thomas C Schleifer, Ph.D

Project Name

Big New Office complex

Completed By | Date

Joe Dimagio | 2023-06-12

Project Size $ | Location (City/State) | Type Of Project

$20,000,000 | Smalltown USA | Commercial

Project Score: 57.5 points

Total Possible: 100 points

Section 1 - Project Fit — Score

1.1 Has the firm successfully completed projects this size (this large)? — 5.0 pts
Never — **Few** — Often

1.2 Has the firm successfully completed projects of this type? — 10.0 pts
Never — Few — **Often**

1.3 Has the firm successfully completed projects in this geographic area? — 5.0 pts
Never — **Few** — Often

1.4 Has firm successfully completed projects with this Owner/GC/CM? — 5.0 pts
Never — **Few** — Often

Section 2 - Firm's Experience — Score

2.1 Has the firm successfully completed projects with this Owner/GC/CM's representatives, inspectors, etc.? — 1.0 pts
Never — **Few** — Often

2.2 Has the firm successfully completed projects with this designer/architect/engineer? — 0.0 pts
Never — Few — Often

2.3 Has the firm successfully completed projects with this designer/architect/engineer's field personnel, inspectors, etc.? — 0.0 pts
Never — Few — Often

2.4 Has the firm's estimating team priced this type of project? — 2.0 pts
Never — **Few** — Often

2.5 Has the firm's anticipated field team (PM, Supt, Foreman) built this type/size of project? — 3.0 pts
Never — **Few** — Often

Figure 13.1 Sample output from the Project Selection Program.

The easiest and surest way to increase profitability is to screen out losing projects, and the Project Selection Program is an innovative, easy-to-use way to do just that. This program is based on the same research that gave the construction industry the concepts of *the contractor of the future, corporate self-analysis, the R-score calculator, financial self-analysis, and flexible overhead*. This program is particularly important now because decisions about which projects to go after are more critical for long-term profitability than ever before. As we have stated in previous chapters, there are no bad projects, just bad matches between projects and contractors. With the Project Selection Program, the potential for a project to make or lose money can be accurately measured before deciding to bid on the project.

The effectiveness of the Project Selection Program is based on the accurate identification of a project being directly associated with the organization's experience with similar work. This chapter will discuss the aspects of a project that most influence performance and exactly how to measure the impact of each aspect. These aspects are identified, measured, and weighted in the Project Selection Program, resulting in a score that indicates how well a prospective project matches a construction organization's experience. The tool works for virtually all members of the industry, including general contractors, construction managers, subcontractors, architects, engineers, and vendors. The tool is particularly useful for project owners and administrators.

13.1 The Origin of the Idea

After decades of identifying and categorizing the causes of poor project performance and contractor failures, we expanded the research to include how to manage project risk and prevent project failure. We wanted to know why so many construction projects failed to produce a profit, even though the contractors undertaking these projects were certain they would profit from the projects. One of the first issues discussed was that, unlike in manufacturing, in which operations improve with repetition, construction projects vary enough that there is not much repetition from one project to the next. Remarkably, the same project may be the best selection for one construction organization and may be a poor fit for another organization. The differentiator is the prior experience of each firm with other similar projects and not the projects themselves.

It turns out that project risk management hinges almost entirely on risk prevention. That is, selecting the right projects and not going after projects that have limited potential for the company to complete successfully. The question should not be, *"Is it a good project?"* After all, every project is good for some company. Rather, the question should be, *"Is this project good for our organization?"* Industry professionals claim to understand this concept, but the number of unprofitable projects is clear evidence that construction professionals either do not understand the principle or, more likely, do not apply it.

Because project risk is based on an organization's prior experience, each new project a contractor considers bidding on must be assessed against past experience. The closer a new project is to previous projects, the more likely the organization will accurately price the work and estimate the performance. Since experience and repetition are key to performance, more experience is superior to less experience. Therefore, no experience with a type of work equates to an extremely low probability of accurately pricing and efficiently executing the work. As previously explained, there are various factors to consider in determining whether a prospective project is similar to a firm's previous projects:

- Size of the project.
- Type of project.
- Geographic location of the project.
- Personnel selected to manage and build the project.
- Features of the project.

While we have already discussed these factors. The following subsections describe their importance for effective project selection and explain how each factor influences the Project Selection Program.

13.1.1 Size of Project

Size refers to the magnitude and price of the project. It is relative to the company's experience, and an average-size project is the size of a project a company typically performs. Most construction enterprises also work on a certain number of both smaller and larger projects. The small projects are typically the most profitable, but there are not enough of them to support the organization. Some contractors have referred to small jobs as nuisance work, others say that small jobs help pay the rent, and some contractors complete them as a service for good clients.

Average-size projects are considered undemanding and even easy projects, in that most estimators in the organization can price them, most superintendents and project managers can build them, and these projects almost always perform as expected. Because of the firm's depth of experience in these projects, the level of confidence in proposing them is high, and the expectation of successfully producing them is equally high. Therefore, the projects are low risk.

Large projects help a company achieve critical mass, support the desire for growth, and provide career advancement opportunities for key employees. Most construction enterprises have fewer large projects than average-size projects because large projects are at the limit of the organization's experience. Top managers usually take an interest in pricing the work on large projects. Even if the firm has previously completed projects of this size successfully, the firm does not have as much experience with these projects as with average-size projects. Therefore, the level of confidence in pricing and completing the work is lower, and the risk is higher.

The desire to build big projects does not automatically translate into the ability to build bigger projects at a profit. If an organization that has been successful in building relatively small projects takes on a much larger project, the organization may have difficulty staffing the project. Further, personnel may have little or no experience with big projects, including more detailed designer inspections and requirements. There will be a substantial increase in the required working capital, and it will be tied up for a longer period than the firm is accustomed to. A change in size increases the magnitude of the risks involved.

13.1.2 Type of Project

As with project size, a firm's prior experience with a project type affects the likelihood that the firm will complete the project at a profit. As an example, a contractor that has been successful in building relatively straightforward warehouses or strip shopping centers, attempts its first complex sewage treatment plant, may be surprised at the differences between the types of projects. The contractor's typical crews and subcontractors may have little or no experience with sewage treatment plants. Pricing unfamiliar work is difficult, and building it would be even more so. This type of project would likely include round concrete structures and all kinds of processing equipment that the contractor would have little to no individual or institutional experience with. The likelihood of successfully pricing and producing the work would be very low, making the project extremely risky.

13.1.3 Project Location

The geographic location of a project is a variable with important implications. Work is often performed differently in different localities for a variety of reasons, including climate and customs. Familiarity with an area is an underrated advantage in the construction industry, and a change in location may have an unexpected learning curve, with differences in how the work is produced, inspected, and paid for. An extreme example of a change in location is a US contractor attempting its first project outside of the country. A less extreme example is a contractor with experience exclusively in rural and suburban areas taking on its first project in a congested, urban area. The contractor will face differences in labor productivity, subcontractor availability, pricing, and local customs and regulations.

13.1.4 Project Team

The team a contractor assigns to a project affects the project risk. For example, the risk increases if individuals who have limited experience with the project type are assigned to the project. The firm may have institutional experience with the project type, but that experience is not imparted to individuals who did not participate in obtaining the experience. However, organizational experience does mitigate risk somewhat: if difficulties arise, the company has the needed experience elsewhere in the organization and can assign that resource to help address the problem.

Personnel risk also depends on whether new hires – who are untested – are assigned to the project. If the crews are composed of both seasoned and new personnel, the seasoned personnel are, at the very least, diluted. If the new personnel were selected because they have superior experience, there is little likelihood that the seasoned personnel can influence performance.

Risk is lower if all team members have experience in at least one of the other factors (project size, type, location, and features). Further, the risk is lower if the individuals have worked together in the past, even if not on the type of project the firm is now considering. Risk is even lower if the team members have worked together on similar projects.

13.1.5 Project Complexity

Unusual project features will also increase risk. Most buildings in the United States are rectangular. The largest portion of roads and highways runs reasonably straight, as do bridges and tunnels. Fewer buildings, roads, bridges, and tunnels are curved. Some projects do have unique features, but most construction firms will have little or no experience constructing these features. Therefore, these projects may have a steep and costly learning curve, amplifying risk.

13.2 Project Selection Program

The factors of project size, type, location, and personnel have a strong statistical correlation with a firm's ability to complete a project profitably. The factors of project features, the project owner, and the contractor's availability to complete the project are also statistically significant, but to a lesser degree. All of these factors are assessed in the Project Selection Program so that contractors can comprehensively measure the risk involved in a project in advance, and therefore determine whether the project is likely to be profitable.

It took years to determine how to measure and weigh these factors in order to produce an accurate project risk score. Issues considered included labor, skill levels, subcontractor quality/availability, estimating, and owner expectations. The program was then tested using hundreds of projects with both successful and failed companies and was able to establish a statistically significant correlation between unprofitable projects and the program's ability to derive scores identifying risk.

By using the tool before pursuing a project, a contractor can know which projects to avoid. The tool can also be used during construction to measure project performance. As a reminder, the Project Selection Program is available at no cost at https://tools.simplarbenchmarking.org/projectselection. A spreadsheet version of the program can be accessed and downloaded using the following link: https://drive.google.com/file/d/1HCAUR1Ewi_9Qf11YztkZdQms-fpwY5l/view?usp=sharing. The spreadsheet version of the program permits users to change the weighting of the questions, customizing it to their firm's experience. We highly encourage assessment of the tools selection criteria, and weighting because they have proven to be statistically significant determinants of successful projects.

The following is a description of the Project Selection Program, explaining the various sections and how to use it most effectively (Figure 13.2). The program guides the user through the process of completing the form with expandable instructions and additional explanations of each question, providing the user with a better understanding of the question   how to most appropriately answer the questions.

Project Selection Program v1.1 copyright Thomas C Schleifer, Ph.D. tschelifer@schleifer.com

If the program user clicks **Introduction**, the following appears:

> There are no bad projects, just bad matches between projects and contractors. The probability of success in go/no-go decisions in project selection can be accurately measured in advance and is directly associated with the construction organization's experience with similar work. The elements impacting success have been weighted to produce a numeric scale of how well the prospective project matches the firm's experience. Designed for General Contractors, Subcontractors, and Construction Managers. Effective for Architects, Engineers, and Vendors; Supports Owner's Contractor Selection.

If the program user clicks **Directions**, the following appears:

> Answer the questions without overthinking them. Total possible scores range from 10% to 90%. The program is based on a firm's knowledge about a potential project. If nothing is known about a project, the probability of success is a "toss up," which in statistical language translates to 50/50 or a 50% probability. Therefore, the program score opens at 50% before adding knowledge about

Section 1 - Project Fit	Weight	*Default Weight*
1.1 Has the firm successfully completed projects of this size (this large)? ○ Never ○ Few ○ Often	**10 pts**	*10 pts*
1.2 Has the firm successfully completed projects of this type? ○ Never ○ Few ○ Often	**10 pts**	*10 pts*
1.3 Has the firm successfully completed projects in this geographic area? ○ Never ○ Few ○ Often	**10 pts**	*10 pts*
1.4 Has firm successfully completed projects with this Owner/GC/CM? ○ Never ○ Few ○ Often	**10 pts**	*10 pts*

Section 2 - Firm's Experience	Weight	*Default Weight*
2.1 Has the firm successfully completed projects with this Owner/GC/CM's representatives, inspectors, etc.? ○ Never ○ Few ○ Often	**2 pts**	*2 pts*
2.2 Has the firm successfully completed projects with this designer/architect/engineer? ○ Never ○ Few ○ Often	**5 pts**	*5 pts*
2.3 Has the firm successfully completed projects with this designer/architect/engineer's field personnel, inspectors, etc.? ○ Never ○ Few ○ Often	**2 pts**	*2 pts*
2.4 Has the firm's estimating team priced this type of project? ○ Never ○ Few ○ Often	**4 pts**	*4 pts*
2.5 Has the firm's anticipated field team (PM, Supt, Foreman) built this type/size of project? ○ Never ○ Few ○ Often	**6 pts**	*6 pts*

Figure 13.2   Sample from the first section of the Project Selection Program spreadsheet.

the project. As knowledge is added, it raises or lowers the score. A higher score represents a higher potential for success, and a lower score reduces potential. There is a benefit to multiple people within an organization scoring projects separately or in collaboration.

If the program user **Clicks on the questions for more detail**, the following appears:

If the user would like more detail or clarification about a question, they can click on the question. Clarification, or the intention of the question, appears.

Note: To provide our readers with a better understanding of the overall program, each question is shown on subsequent pages, immediately followed by the clarification that appears in a dialog box if the program user clicks on that question.

If the program user clicks on **Change weights**, the following appears:

> The weighting of each question is derived from 40 years of measurement and experience and is considered accurate. While it is not recommended, users may change the weight for any question they feel a different weight may be more appropriate for their circumstances. The **default weights** are shown in the far-right column to remind a user who has changed a weight how close or far the changed weight is from the program weight. Weighting is based on statistical analysis, which takes into account the likelihood that an event will occur and the severity if it does occur. For example, if a costly issue (severity) occurs historically in a small percentage of projects, it is weighted lower than an issue with high severity and high occurrence. Users should change weighting values carefully, keeping in mind that varying one point on a 10-point question revises the weighting by 10%, while varying one point to a one-point question revises it by 100%.

13.3   Program Questions

The following are all of the questions included in the program. Each question is followed by the clarification or reason for the question that would appear in the dialog box if the program user clicked on that question.

Section 1 – Project Fit

1.1	Has the firm successfully completed projects this size (this large)?	
	Attempting projects larger than a firm has experience with, generally measured in dollars, is a discriminator of project success. A project 10% larger than anything the organization has profitable experience with is reasonably low risk. A project twice as large is high risk. The user should use judgement in pursuing projects larger than they have successful experience with.	10 pts
1.2	Has the firm successfully completed projects of this type?	
	Similar type of project means work that is comparable in nature to work the organization has specific experience completing at a profit either with their own forces, subcontracted or a combination of both. Type of project includes contracting method such as CMAR, Design-Build, Design-Bid-Build, etc.	10 pts
1.3	Has the firm successfully completed projects in this geographic area?	
	Geographic area is a significant discriminator of potential project success because of the impact of local knowledge and differences in local practices. This question addresses the organization's experience in the area of the project. Has the firm worked successfully and profitably here under current management?	10 pts
1.4	Has firm successfully completed projects with this Owner/GC/CM?	
	The owner is the entity that will own and/or pay for the project and may or may not be the direct user of the project. For example, a state may procure a road project (owner), but not be the user which is the public. The question addresses the entity that the user will contract with and answer to and if that is two separate parties or agencies both should be considered.	10 pts

Section 2 – Firm's Experience

2.1	Has the firm successfully completed projects with this Owner/GC/CM's representatives, inspectors, etc.? The owner's representatives, inspectors, etc., is the person or team the owner delegates to represent the owner's interests and carry out owner responsibilities as designated in the contract. This may include items such as approving change orders, quality inspections, payments to contractors, quality control, etc.	2 pts
2.2	Has the firm successfully completed projects with this designer/architect/engineer? Prior experience with a project designer, architect or engineer is a significant indicator of a potential successful relationship and includes knowledge about the level of detail and clarity of design documents, type of cooperation the designers offers, the buildability of their designs, etc. Lack of direct experience may be compensated for by knowledge of a designer's reputation, however the user is cautioned that such information is difficult to verify and may have limited reliability.	5 pts
2.3	Has the firm successfully completed projects with this designer/architect/engineer's field personnel, inspectors, etc.? Has the firm had successful experience with the designer's, architect's or engineer's field representatives, inspectors, and other who may represent the designer's interests and carry our designer responsibilities? These are usually designated in the contract which may include, but are not limited to: items such as issuing change orders, quality inspections, approving payments, the speed with which requests for information and shop drawings are responded to, etc.	2 pts
2.4	Has the firm's estimating team priced this type of project? What is the level of experience of the firm's estimators with similar types of work including similar materials, subcontractors, vendors and labor? The user should take into consideration those who will be preforming the actual estimating for this project.	4 pts
2.5	Has the firm's anticipated field team (PM, Supt, Foreman) built this type/size of project? The likelihood of success is impacted if the field team that will be responsible to construct this project had direct experience building and/or managing this type/size of project. This does not mean "years in construction" but direct hands-on experience with similar type of work.	6 pts

Section 3 – Project Complexity

3.1	Is the required project schedule reasonable? From the experience of the person filling out this form is the duration of the project defined in the contract reasonable. If a duration or completion date is not defined, or the person filling out the form cannot determine if the duration is reasonable, the answer is "unknown".	4 pts
3.2	Is there sufficient labor available to produce the project? (consider available skill levels) Is labor availability in the area where the project will be built and at the time the project will be constructed going to impact production or schedule? Labor skill level should also be considered. If uncertain answer "unknown". Achieving unreasonable schedules impacts cost and not achieving them impacts reputation.	5 pts

(Continued)

(Continued)

3.3	Has the firm had prior successful experience with the intended major subcontractors/vendors/etc.?		5 pts
	This addresses the level of confidence the firm has in the major subcontractors that will be awarded work on this project. Successful prior experience is a positive as is reputation, however if there are a large number of subcontractors this question is difficult which in itself highlights concerns about unknowns. The less known about subcontractors' performance the greater the exposure.		
3.4	Are there unusual access, room-to-work, parking, dust, noise, storage space, traffic, etc. issues?		3 pts
	This issue address convenience and ease of operation which can have a huge impact on productivity. For example, an inner city project with zero storage space is managed differently that an open-field project just as working in occupied spaces has different exposures and requirements that unoccupied spaces. Access, room-to-work, parking, dust, noise, storage space, traffic issues or others have associated costs and productivity impacts.		
3.5	Are there unusual ground or surface water conditions?		3 pts
	Obvious and announced ground or water condition can be difficult to estimate which impacts costs and suspecting such conditions does the same. Prior experience in a particular area may offer reasonable knowledge about anticipated conditions which may reduce exposure. If there are no obvious or announced conditions the answered is no. Unknown is a relatively common answer but would be recommended only if the user had reason to suspect or some doubts about the issue.		
3.6	Are there rock, soft soils, unusually deep excavations, etc. that may complicate or delay the project?		3 pts
	Obvious and announced rock, soil issues or deep excavations may be difficult to estimate which impacts potential success and just suspecting such conditions does the same. Prior experience in an area provides superior knowledge about anticipated conditions and reduces risk. Unknowns increase risk.		
3.7	Are there unusual, circular, curved or angled design elements: i.e. walls, bridges, roofs, structures or other that may complicate/delay the project?		4 pts
	Straight lines, squares and rectangles are most common in the built environment and therefore designers and constructors have more experience with them. Conversely fewer have experience with round, curved or oblique angles. A project with unusual, circular, curved or angled design elements: i.e. walls, bridges, roofs, structures, or other may be more complicate than anticipated and/or delayed. Users may want to take into consideration that out-of-the-ordinary designs may impact estimating and/or production in measuring potential success. If the firm has direct prior experience with such issues potential for success is increased.		
3.8	Are there critical long-lead, sole source or unusual items that may delay the project?		2 pts
	Critical long-lead, unusual or sole source items can seriously impact schedule. At project selection phase little may be known about how long it will take to procure the items so potential success is impacted. Unknowns increase concern and knowledge reduces it. However, because the user does not have the job yet clarifying these issues may not be practical. In which case caution should be exercised.		

Section 4 – Financial and Cash Flow Impacts

4.1	Is the project owner's financing in place?		
	There is exposure associated with pursuing a project when the owner's financing is not firmly in place, however, the exposure at pre-project stage is limited to the cost of pursuing the project. Pre-project potential is being measured up to, but not including the award of the contract so it is assumed that without financing in place most firm's would not enter into a contract. Some may not expect all financing to be firm at this stage and may base their pursuit on the assumption that it will be by contract award.		2 pts
4.2	Does the firm have experience with the payment processes of this owner/GC/CM?		
	Project owners vary considerably in payment processes. How long it takes to be paid obviously impacts cash flow and has a cost associated with it which may affect project selection decisions. Experience with an owner's payment process (good or bad) moves the issue of payment to known which reduces exposure. If a user likes or dislikes the process they can base their pricing and decision on experience. Conversely having no knowledge about an owner's payment process increases exposure at the project selection stage.		2 pts
4.3	Will this project strain the organization's resources: i.e. financial, equipment, manpower, management, or other?		
	An organization's resources include, but are not limited to, manpower, equipment, financial, management or other. If the firm's resources are fully committed and outside resources are not available, profit potential will be impacted. Resources from within the firm are known and therefore easier to deploy. If resources must come from outside the firm efficiency and cost may be impacted.		3 pts

Section 5 – Contract Issues

5.1	Are contract terms reasonable?		
	Contract terms impact performance and determining the reasonableness of contract terms is difficult at best. Because most contracts are issued to the firm by the owner, negotiating terms may be difficult to impossible. A firm may need to determine what contract language it is willing to accept and will the cost for any exposure be included in the estimate? This will often be answered "unknown" as contract terms may not be available when a project selection is being made.		1 pt
5.2	Are there any unusual payment provisions, retainage issues or liquidated damages that will affect cash flow?		
	Retainage is a cash flow issue that most firms understand and have dealt with in the past. This question addresses any "unusual" retainage issues that should be considered at pre-project selection stage with an understanding that retainage impacts cash flow.		1 pt
5.3	Are there any unusual bonding requirements?		
	Bonds obviously have a cost and may impact on a firm's bonding capacity. Unusual or out-of-the-ordinary bond language is becoming more prevalent and can seriously impact project potential.		1 pt

(Continued)

(Continued)

5.4	Are there any unusual insurance requirements?	
	Some contracts require the contractor to provide insurance for issues that are difficult or expensive to insure and some have requirements to insure third parties. A contract that requires no additional insurance other than what a firm already caries has low or reasonable exposure. Contracts that require additional or special insurance or coverage for third parties has greater cost and exposure.	1 pt
5.5	Are there any unusual indemnity issues?	
	Some contracts require the firm to indemnify others, which could expose the firm to becoming responsible for the actions of others which may have serious cost consequences. With no unusual indemnity issues or only those the firm has undertaken in the past many consider the exposure to be "normal" which for these purposes would mean low. Unusual or increased indemnity issues in a contract increase exposure and should be considered in project selection decisions.	1 pt

Project Fit Issues

These are six project issues that can impact project selections, but not consistently enough to have been in the above questions. They are included after the program questions to remind the user that these issues should be considered in any project selection. These issues are particularly appropriate for those selecting the project to discuss as a group.

Section 6 – Additional Issues to Consider (Not Scored)

6.1	Do you believe you have a competitive edge on this project?
	Going after a project generally has a cost associated with it along with potential lost opportunity if the use of resources to get the project prevents or impacts going after another job or reduces efforts of attracting other projects. This makes the measurement of a competitive edge or lack of one significant. Where there is a competitive edge on a project it may make sense to invest more resources than on one where there is less or no competitive edge. This is difficult to measure, but for most contractors it is in intuitive.
6.2	Do you believe another contractor(s) has a competitive edge on this project?
	Realistically recognizing any competitive edge other contractors may have is valuable knowledge and can assist in a project selection decision. Most are familiar with projects where a favored contactor was the owner/GC/CM's obvious choice, and many other contractors wasted a lot of effort and resources only to see the obvious choice selected. Screening out projects that a firm has limited chance of capturing saves considerable resources which can be applied to projects with a higher likelihood of success and can potentially increase sales.
6.3	Is there consensus within your organization about perusing this project?
	When everyone in an organization is excited about pursuing a particular project the odds of capturing that work go up through the quantity and quality of the energy expended. The converse is true when there is little enthusiasm for a project making capturing it more difficult. There are situations when not going after a project is the best project selection decision. If some or many are concerned about a certain project it may be the wrong job for the company.

	Is your "gut feeling" about this project negative or positive?
6.4	The expression: "If all else fails trust your gut" can be a valuable inclusion in the project selection process. The information collected from the above questions is of great value, however, if persons in responsible positions within an organization still have a gut feeling about a positive or negative decision it probably makes sense to reevaluate the data before making a final decision.
	How does the project fit with the existing/anticipated backlog?
6.5	A contractor whose current work is running smoothly has only limited exposure in taking on additional work. However, a contractor managing a troubled job or two may encounter additional risk in adding the next project. A well-managed and organized backlog has reasonably low exposure and makes additional, prospective projects attractive. A troubled backlog has higher risk which may indicate that this may not be the time to add another project.
	Does the contract include Consequential Damages or are they expressed or implied?
6.6	Consequential Damages are damages resulting as a consequence of the project. The owner is the usual party damaged, the public and others may also be included. Fairly common are damages that result when a revenue-generating project is delivered late such as a restaurant, hotel or retail business. The damages include loss of revenue because the intended use is delayed which can result in serious potential liability for a contractor that causes this to happen. Some contractors will not undertake such exposure which unlike liquidated damages cannot be measured in advance.

After the description above, there is a further explanation and clarification about how and why the program questions are weighted and the reason that readers have the option to alter the weighting (which is not recommended).

All of the questions are listed, followed by the question clarification that would appear if the reader clicked on the question.

To familiarize readers with the tool, we will walk through its components. The tool's webpage begins with an introduction to the tool and then provides the following description of how to use the tool:

> Answer the questions without overthinking them. Total possible scores range from 10% to 90%. The program is based on a firm's knowledge about a potential project. If nothing is known about a project, probability of success is a "toss-up," which in statistical language translates to 50/50 or a 50% probability. Therefore, the program score opens at 50% before adding knowledge about the project. As knowledge is added, it raises or lowers the score. A higher score represents a higher potential for success and a lower score reduced potential.

There is a benefit for the company to have multiple people within the organization score projects separately and review them as a team.

Following this description, there are fields in which to type the project name, current date, project completion date, project size (in dollars), project location, and project type. Next is Section 1, which asks questions to gauge the project fit. Next to each question is a question mark icon that users can hover over to see more information pertaining to the question.

If the program user **Clicks on the questions for more detail**, the following appears:

> If the user would like more detail or clarification about a question they can click on the question. Clarification, or the intention of the question appears.
>
> ***Note:*** To provide our readers with a better understanding of the overall program, each question is shown on subsequent pages, immediately followed by the clarification that appears in a dialog box if the program user clicks on that question.

13.4 Scoring a Project

To the right of the questions, there is a column heading "Weight" which is the weight given to each question. As mentioned above, the answers to the questions vary in their importance or significance, which in turn affects how much they influence a project. Each topic is meaningful and therefore important, but some are more important or have a greater influence on project outcome or risk than others. Using data collected from thousands of projects and the impact the outcome had on a company's performance, the authors were able to calculate the variance in their significance and the weighting of each was a measure of those findings. The scale used was 1–10, with 10 being the greater significance and 1 being the least significance.

The program weighs each question, and each question is assigned a default number of points. Permitting the default weights to be changed was a very controversial issue during the development of this tool. About half of the research team was opposed to offering the users the opportunity to change the scientifically arrived-at weights. Most of the other team members believed that some construction professionals feel strongly that they know best about their businesses, and if they thought the weight given to a particular item was incorrect, they might not use the tool. The lead researcher broke the tie in favor of offering the ability to change the weights. Consequently, "Change weights: click here" was included in the tool.

By examining data from thousands of projects, we were able to calculate how strongly correlated each factor is with project risk. We then weighted these factors in the Project Selection Program according to their level of significance. The scale ranged from 1 to 10, with 1 being the least significant and 10 being the most significant.

The tool starts with the assumption that the company knows nothing about the project. In this case, the tool assumes a 50% probability that the proposed project matches an organization's experience. This percentage also indicates that the project has a 50% probability of being successful for the company. As a person answers the questions in the tool, the probability changes. For example, the first question is, "Has the firm successfully completed projects of this size (this large)?" If a person indicates that the firm has not, the probability (or risk) that the project is not a good fit for the company increases. If the firm has often completed projects of the same size, the risk decreases. As another example, if the project is a type of construction that the organization has never attempted before, the probability increases. If the organization has regularly performed this type of work, the probability decreases. If the contractor has completed this type of work only a few times, the probability remains unchanged. Answering the questions will require judgment, which an individual with full knowledge of the organization is equipped to make. If a question is not answered, the probability remains 50%. Obviously, unknowns exist, but it is preferable to leave as few unknowns as possible, but from a statistical perspective in the tool, the probability is 50/50.

It should be apparent that no organization should take on a project with only a 50% probability of success. A project with a probability lower than 50% should be out of the question no matter how desperately a firm needs the work. (No firm ever needs a loss.) Each contractor must decide the lowest percentage probability the contractor is okay with, but anything lower than 75% is not a good option for most contractors.

As information becomes known (knowledge), it alters the probabilities (odds). For example, if the project turns out to be much larger than the contractor has ever attempted, the probability, or the risk, that it is not right for the company goes up. If the size is right, the risk for the organization goes down. If nothing is known at any point in time, the risks or odds are unchanged and remain 50%. Similarly, if the project is a type of construction that the contractor has never attempted before the risk increases, and if the contractor has regularly performed this type of work, the risk decreases. If the firm only did this work, a few times the risk is unchanged. These answers require judgment, which a user, who has full knowledge of the organization in question, is equipped to make.

At varying points in time, some important information will be known, and some not. As knowledge increases, the risk alters up or down. For questions where nothing is known, the "unknown" does not statistically alter the risk, which for that line item remains 50%. Obviously, unknowns exist; however, it is a statistical draw – no change. The program is based on the odds of no knowledge being 50/50 and as knowledge increases the odds alter up or down. The major categories that impact project risk have been presented earlier and as a reminder are discussed briefly in the next section. It should be apparent that no contractor should risk a project with only a 50% probability of success. A probability of lower than 50% should be out of the question no matter how desperately the firm needs the work. (No firm ever needs a loss.) Each of us picks the odds we are willing to take; however, anything less than 75% does not work for most businesspeople. There is no 100% probability of success, so the amount of risk remains up to the contractor.

To summarize our discussion about project selection risk, we find it most appropriate to repeat the next several topics.

13.4.1 Project Size

Construction organizations produce projects of varying sizes. Size, of course, is relative. A large project for one firm may be a small project for another firm. *Small* and *large* are used here in relation to a company's average size of projects. Our experience indicates that companies typically complete fewer small projects than average and large projects, though small projects typically earn the highest profit percentages. Average-size projects are generally profitable but at a lower percentage than small projects. Large projects are usually few in number and earn a smaller profit percentage than average-size projects do.

Small projects are often performed as a service for good clients. Though these projects typically yield the highest profit percentage, there are not enough of them to support the organization. Some contractors refer to these projects as nuisance work, but most contractors admit that small jobs *help pay the rent*. Average-size projects are often the main source of profit and cash flow and are therefore the projects a company relies on and survives on. Average-size projects can be described as the company's *undemanding*, even easy projects, in that most estimators in the organization can price these projects, most superintendents and project managers can build these projects, and these projects almost always perform as expected. Therefore, a company has great confidence with average-size projects, and the risk is low.

Large projects can help a company meet critical mass. These projects support the desire for growth and appeal to the interests and ambitions of key employees. Large projects differ from average-size projects in that not all estimators can price large projects accurately, top managers take an interest in preparing the estimate, and there may be long hours or days of work before the bid is submitted. The greater concern about pricing a large project is indicative of the greater risk. What may not be as clearly understood is that greater attention is also a sign of concern because of limited experience with large projects. Even if the company has successfully completed projects of the same size in the past, the company has not completed as many of these projects as it has completed average-size projects. Therefore, there is less cause for a high level of confidence. The amount of experience with any project correlates with the level of confidence in pricing and expectation of success.

It should also be noted that project size affects the risk on a monetary scale. Small projects that do not do well or lose money are not large enough dollar amounts to have a financial impact on the company. An average-size project that loses will hurt the company but generally will not cause the company to fail, because a number of other average-size projects usually take up the slack. In contrast, a large project that underperforms or loses money has the potential to materially affect the company's financial condition.

13.4.2 Project Type

As with project size, prior experience with the project type correlates with the likelihood of successfully pricing the project and producing it on time and on budget. The likelihood of successfully pricing and producing a new type of project is low, resulting in high risk. If most or all of a firm's prior projects are the same type of work as the new project being considered, the type of work will not increase risk.

13.4.3 Geographic Area

Because construction work is produced differently in different geographic areas, experience working in an area affects the likelihood of success and thus affects the amount of risk to the contractor. An extreme example is a contractor located in the Arizona desert attempting a project in Minnesota during the winter. A less extreme example is a contractor with exclusive experience in rural and suburban areas taking its first project in a large urban area. The project's working condition may be outside the organization's experience and would therefore present a learning curve. The likelihood of success would be lowered, presenting a risk to profitability.

13.4.4 Project Team

As previously explained, if project team members do not have direct experience working on a similar project, then the risk will be higher than if one or more team members have prior experience with the project size, type, location, and so forth.

13.4.5 Unusual Project Features

Most buildings in the United States are rectangular. Most roads and highways run relatively straight for much of their length, and bridges and tunnels are straight to the extent possible. Therefore, the lion's share of experience collected by construction enterprises is with straight lines.

If a potential project has a curved wall, windows, or roof or other unique elements, this project is outside the experience of most organizations. If an organization does have such experience, the odds are that the experience is limited. The same goes for out-of-the-ordinary roads, highways, bridges, and other types of projects. Because of a lack of experience with these types of projects, there is considerable risk in proposing and producing them.

One might say that a totally unique project is risky in and of itself and that each firm that attempted it would be equally at risk. This assumption is not accurate. The contractor that has the closest experience to the unique project would have risk, but the risk would be smaller than for a firm with less-close experience. If no contractor had any even somewhat-close experience, then the risk would depend solely on the project size as measured in dollars. The contractor for which the project was the smallest relative to the contractor's average-size projects would be at least risk. This is because if the project failed and generated a loss, the loss would have a smaller effect on the contractor than the loss would on a contractor whose average-size project was smaller than the unique project.

We need to emphasize that none of these projects have intrinsic risk. Project risk is exclusively a result of an organization's experience with similar projects. Because an organization's experience resides in its employees, selecting the right personnel for a project is critical. To reiterate what we've previously stated, project risk can be measured in advance. We developed a tool to measure the risk. The Project Selection Program accurately measures project risk to a specific enterprise.

13.5 Conclusion

Our research confirmed the need for a process to screen out high-risk projects during the opportunity stage of pursuing work and attempting to minimize unprofitable work. The reason we spent an additional year and a half conducting research and analysis to develop such a process: the Project Selection Program.

During testing of the program, numerous contractors applied the tool to three of their completed projects: two successful projects and one project the firm wished it had not taken. Each firm answered 26 individually weighted questions, and the resulting score served as a measurement of how the firm's actual experience aligned with the tools' predictions for the three jobs. The results demonstrated that the tool predicted the actual outcomes 100% of the time. The tool projected that the successful projects would indeed be successful and that the losing project would be unprofitable. Extensive testing validates the tool as an effective means of predicting the potential success of a project. The Project Selection Program is an easy-to-use tool that has been placed in the public domain, available at no cost: https://tools.simplarbenchmarking.org/projectselection. Users are encouraged to complete the same type of three-project test to gain confidence in using the tool as an effective method for screening potential projects.

14

Project Controls

Most of the work in the construction industry is obtained through competitive bidding or some form of estimated anticipated costs prior to construction. Bidding or pricing a job is accomplished by estimating the cost of the work to establish a bid or budget price. Estimating the cost of the work before performing the work also establishes the project's budget, providing contractors with a detailed financial plan to use when executing the project. Unlike many other industries, a detailed estimated cost to perform the work is inherent in the contractor's process for obtaining their work. In some cases, the budget is overlooked or underutilized as a tool for tracking progress and managing work as it progresses. Once a project starts, material hits the ground, trades begin working, and the clock starts ticking on the limited time given to complete the project.

The launching of a project requires so much effort that it is hard for project teams to find time to develop and implement the detailed financial controls that are vital to effectively managing projects. *Control*, as we use the term here, refers to the project team's ability to quantify the work and measure how the project is progressing relative to the plan and schedule. As mentioned earlier, the authors prefer the term "manage" rather than "control" because while we attempt to control progress all we can really do is try, which we consider "managing" the project. As "control" is so widely used in this context, the authors use the term control in this chapter as interchangeable with "manage." An effective control system not only enables contractors to know how they are doing, but to predict how much time and money it will take to complete the remaining work on the project and to use that information to make informed decisions to manage the project.

For most companies, a project manager is entrusted with the responsibility of ensuring a project is completed on time and under budget. The project manager is essentially the CEO of their own specialized group organized for the purpose of building the project, which starts with the execution of the contract and ends with the completion and acceptance of the project. A construction company is merely a collection of all the projects it has under contract, and each project acts as a profit center contributing to or detracting from the company's bottom line. A construction company's success is ultimately determined by the total profit or loss collectively realized by all the individual projects the company builds, less its overhead costs of doing business. Construction projects are rarely successful without a good project manager, but a good project manager alone is not enough. The project manager's success is often based on the processes and systems of controls put in place by the organization for managing projects.

The Business of Construction Contracting: Schleifer's Guide to Financial Success, First Edition. Thomas C. Schleifer and Aaron B. Cohen.

Most think of financial controls as being synonymous with accounting because it's easy to relate the tracking of project costs to an accounting function. Accounting is intended to keep score and measure the financial health and performance of the company as a whole. It effectively tells the story of how a company has performed financially in the past, but can be limited when it comes to predicting how a construction company will perform financially in the future. This is particularly true for a company where the vast majority of its costs are determined by the management decisions made on the projects it performs. This is especially the case of companies whose backlog consists of a fewer number of large-sized contracts. The types of historical trends that can be effectively used in other industries to predict future performance do very little to offer predictions on how construction companies will perform on a project-by-project basis because such historical trends do not accurately represent the complexities involved in completing the work in progress (WIP).

Accounting metrics are effective at indicating things like how much net worth the company has or when it will run out of cash at a given rate of spending. Accounting measures like liquidity and net worth are of great importance to banks or bonding companies, and by extension should be of great concern to contractors because they explain the company's perceived financial health. However, such accounting metrics do little to help contractors run their work and effectively manage the costs to complete work under contract. Effectively controlling project costs requires assessing progress against the planned cost of the work and identifying variances from the plan early enough to do something about it. Most accounting systems are simply not equipped to explain things like why the last 10% of the work on a project ended up costing 20% of the project budget.

Project controls include the project management methodology used to track the progress of work against plan, identify variances, and support the management decisions needed to guide a project to the achievement of its goals of finishing on time and within budget. The construction process is complex, requiring the effort of multiple specialized trades working together to build a complex structure of inter-related systems in a compressed time frame. Because the success of a construction company is predicated on the performance of the projects, it is critical to measure progress against that plan to identify deviations in specific areas early enough in the process to take corrective actions and maximize the likelihood of keeping the project on track. As a company grows and more individuals become responsible for making project-level management decisions, the need to have systems of control in place to ensure projects are being effectively managed becomes increasingly vital to the success of a construction business.

14.1 The Evolution of a Construction Business

When a company is small, it is typically successful because there is a single person (or few) who excels at building the work. As a company grows, it tends to be constrained by the individual's ability to be involved in every major decision in the management of all projects. When more work is obtained and projects are running concurrently, the attention of the individual is spread across multiple projects, and it starts to become even more difficult for a single person to be everywhere at once and to remain in control of all the work. As the company continues to grow and when the projects become larger and more complex, the need to delegate decision-making responsibilities to others becomes paramount. To ensure continuity and consistency, systems need to be put in place to guide the increasing number of managers involved in the work. These systems and procedures are necessary to enable less experienced, but proven and capable individuals to assume additional responsibility while still enabling the key individuals of the company to retain a level of understanding of the status and progress of the work.

The concept of implementing project control systems can be challenging for the founder of a company whose success was predicated on their entrepreneurial ability to overcome all manner of never-ending obstacles threatening their business. The very notion of delegating control to other, less-experienced individuals and relying on processes and procedures to ensure the profitable completion of projects may go against every instinct of the founder. Unless a company can figure out how to overcome the challenge of a key person being everywhere at all times, it cannot grow beyond the level of whatever volume of work the key individual can handle personally.

Organizations that lack the systems necessary to delegate responsibility to individuals or project teams tend to stagnate at some point in their growth. Making the transition from a company known for the capabilities of a few key individuals to being known for its ability to produce quality work consistently across various types, sizes, and locations of projects can be a difficult transition. To accomplish this requires a system of controls to enable the delegation of authority and responsibility for decision-making to senior managers. This includes processes and procedures that accurately and timely report to management how the work is progressing. The performance of the projects must also be accurately reflected in the financial reports of the company, which enables senior management to make informed decisions about the direction and growth of the business.

14.2 Construction Is a Service Business

In business, success is defined by a company's ability to produce a return on investment for its shareholders. Shareholders may be a collection of investors and equity partners, or they can be a single company founder who owns 100% of the company, but in either case, there is an investment of capital in the company with an expectation of the company producing an acceptable return at some point in the future. Success, for most, is defined as producing profitable work and increasing net worth.

It is useful to compare the differences between a service business, such as construction, and a more conventional type of business, such as manufacturing. For a company that produces a product for sale, it is common to test a market to establish a need or demand for the product the company is capable of producing. If there is, a business plan can be established with the intent of attracting investors that would provide capital. Investors assess the risk and invest according to their expectation of earning a return on their investment. With a target market identified and capital secured, the investment in plant and equipment is made to establish a manufacturing operation. The operation, which is capable of converting raw materials into finished goods, produces products that are then sold into the market. As production is increased, the cost to produce each unit is reduced and excess production can be stored as inventory, which becomes an asset of the company. Sales goals are established, and costs are closely monitored with the intention of producing and selling the products at a profit and, over time, increasing the net worth of the organization. The entire process is closely monitored and controlled, ensuring costs are managed effectively and the amount produced matches the demand in the market.

Some classify the process of construction as a mobile manufacturing operation, but that is not correct. Construction is a service business where the true assets of the company reside in the collective experience of its employees' abilities to build certain types of work. So, unlike a manufacturing business, where product lines can exist for years and the production cycles can be highly repetitive, construction projects are hardly ever the same, and the economies of scale realizable through repetitiveness are largely unattainable. Even where projects are identical, the uniqueness of existing conditions across various project sites can impose so much variability in the work that the same project on a different site can essentially be an entirely different type of project.

One example would be installing a waterline in rock instead of soil. Digging through soil requires a contractor that has an appropriately sized excavator and experience in the work. However, for the same waterline to be installed in rock conditions requires a completely different type of contractor that is able to assess the hardness of the rock and is equipped to perform the necessary drilling and blasting of the rock as well as possessing the skills necessary to install the waterline itself. The experience required to perform such work may result in a limited number of qualified contractors even available to bid on the work, which drives prices higher due to limited options. The availability of qualified contractors may even be a controlling factor in when the work can be performed instead of the project's schedule. From a design perspective, these two jobs could have identical requirements, but from the contractor's perspective, these are two completely different jobs.

Construction is a service business where the service of delivering projects on time and within budget is dictated by the people and processes employed to provide that service. Training and retaining key people are a crucial part of that capability, but good people alone cannot ensure the quality and efficiency needed for a company to produce work at scale. The *processes* a company employs to build a project provide a competitive edge to those with experience who have built in consistency and repeatability to produce work better and faster than their competitors. The following sections are explanations of the various tools and processes available to contractors with the capabilities necessary to effectively *control their projects*.

14.3 Productivity Analysis

One of the most important metrics for contractors to monitor and control is *productivity*. Productivity is the rate of output relative to a given input. It is a measure of efficiency and is typically expressed as the number of units produced per measure of time, such as 30 cubic yards of concrete placed per hour, or 200 linear feet of pipe installed per day. While factory-like production rates can be precisely honed and improved over time, such efficiency improvements are not practical in construction. Construction professionals need to monitor production and track productivity relative to a plan to measure and understand how well they are performing.

All projects have a budget and a schedule, and whether explicitly used to manage a project or not, production rates drive both. If production rate targets are not specifically stated, they are at least inferred based upon the estimated cost to produce the work, which, if exceeded, will result in costs that exceed the budgeted amount the workforce is required to maintain in order to complete the work for a profit.

Companies employing their own tradespeople are commonly accustomed to managing work by closely monitoring productivity rates. Labor productivity is the single most volatile cost and largest risk factor for any construction project. This is because the production rates presumed when pricing the work cannot be known to any finite degree of certainty. In a factory, the conditions are highly controllable and predictable, while on a construction project, the variables that can influence productivity change from

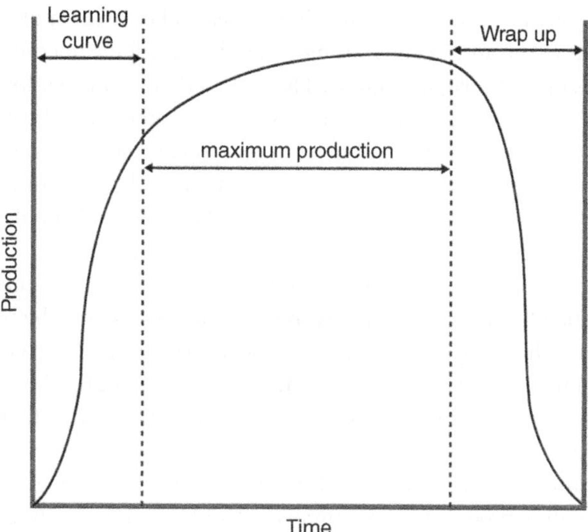

Figure 14.1 Typical production learning curve graphed over time.

job to job, site to site, and even from crew to crew based on the individuals performing the work. When work is performed by subcontractors, often in an attempt to shift the productivity risk to a more experienced party, the contractor still needs to monitor the work and provide support in order to achieve optimal output. Production rate slippages obviously impact the overall project schedule.

When analyzing productivity, it is important to consider the learning curve that most construction tasks will experience. At the beginning of a project, the crew gets accustomed to the work and learns the process that works best for a particular project. Production typically ramps up to peak efficiency, and then trails back down toward the conclusion of the activity (Figure 14.1). The completion of the work is also where high output is much more difficult to attain.

Productivity in the field is largely influenced by management. A general contractor's control of the site drives the logistics of how the work is laid out and how subcontractors and suppliers can access the site and perform their work. Ensuring access to the work is critical and easier said than done. Particularly when securing the site is also necessary. Reducing the distance between a staging area and the location where the work is performed can have a dramatic impact on the productivity of crews. Management provides the tools for workmen, training, and direction on what work is required by when, and has the ultimate obligation to ensure the people working on the site have everything they need to perform their work as efficiently as possible. A company's culture, such as attitudes toward providing safe work environments, has proven to have a significant influence on productivity.

Experienced estimators understand the productivity rates for the various scopes of work their company performs, which they use in producing an estimate. The productivity rates are derived from the cost and duration of various past work activities accomplished, and some may be derived from a production tracking system used on prior work. Some companies may be reluctant to share cost information with field personnel, which, in the authors' opinion, is at best an old-fashioned idea and at worst a huge mistake. Tradespeople today are better educated than in the past and most are motivated to perform their trade to the best of their ability. The authors feel strongly that the workforce not only wants to know what is expected of them, but also has the right to know. Those directly responsible for producing the work need to understand when they are doing well. They also know that producing the work for a profit is the ultimate goal of the company they work for and most expect to participate in achieving that goal. The workforce of today almost expects to participate in a system that tracks the work based on production targets. Productivity targets, of course, need to be reasonably attainable and should stretch the capabilities of a crew, but not be so aggressive that the crew feels they are unattainable.

Another consideration in monitoring productivity rates is the use of the information. At the conclusion of the project, it's best practice to conduct a post-mortem and assess how the project went. Material and labor costs are constantly changing from project to project, so looking at unit costs for materials may be of limited value. Reviewing productivity rates for similar work is an appropriate comparison, even for projects that were completed years apart. Because there are so many things that impact production, *benchmarking*, or performing comparisons between different jobs can be difficult, but certainly worth the effort (Figure 14.2). Failure to do this type of analysis at the end of each project allows faulty assumptions made during the estimating process to go undiscovered and be continued into future estimates and project budgets. Building a database of production rates is an incredibly valuable resource for estimates of future work and is a primary goal of a learning organization. It provides a system for monitoring and improving productivity from job to job.

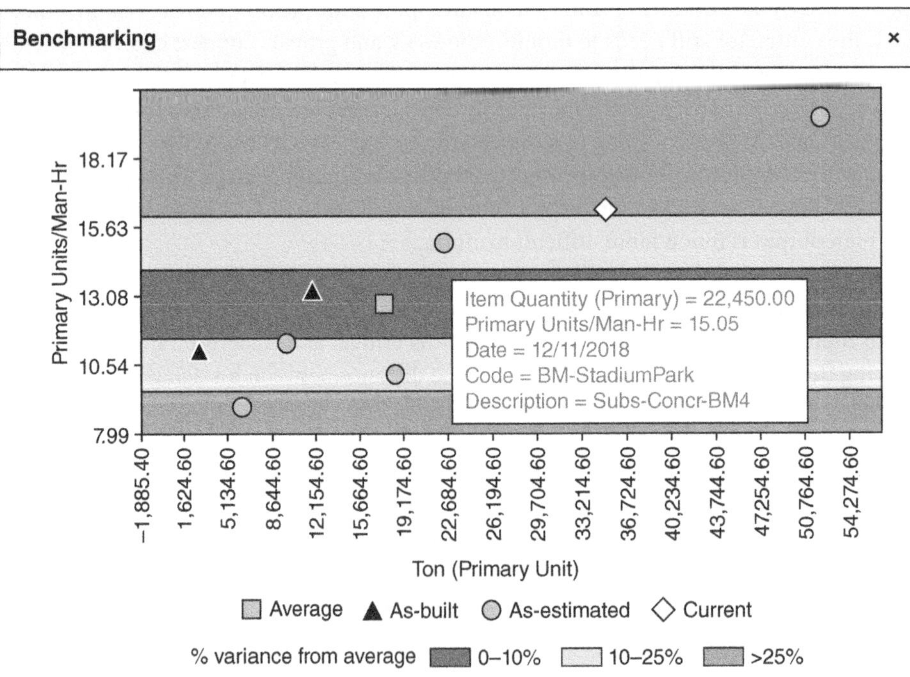

Figure 14.2 Example of a benchmarking application.

14.4 Schedule

The progress schedule is one of the most essential tools in managing a project. Most organizations use the project schedule as a planning tool to understand the sequence of work and how various activities relate to each other. With relationships between activities understood, dates can be established for the various tasks required, and the overall project duration can be accurately determined. The schedule becomes the primary tool for tracking the progress of the work as it progresses, for comparison between planned and actual durations and dates. Typically, creating the schedule for a project is done with a software application that automates much of the work (Figure 14.3). The schedule represents the work plan and timing. The scheduling software streamlines the process of updating assumptions made and changes to the work plan, activities, dates, or durations. As updates are made in response to modifications to related activities, the impacts on the overall project duration can be quickly understood. The process also facilitates brainstorming of different approaches to building the work by creating different versions of the schedule. This assists in the identification of optimal work plans.

Schedules, specifically Gantt charts, are highly visible in nature and are a great way to communicate a lot of information in a concise format. Accordingly, they are excellent visual tools to use with owners, designers, or other external project stakeholders that don't typically have a particular interest in the details of the work but have a high interest in the overall duration of the project. Users of a schedule can easily see how long a project will take and what the long-duration activities are. The schedule also indicates the critical path, or which particular sequence of activities depending on one another will result in a delay to the overall project schedule if a delay occurs on any one of those individual activities. Activities not on a critical path are said to have float, which means there is more flexibility in their start and finish

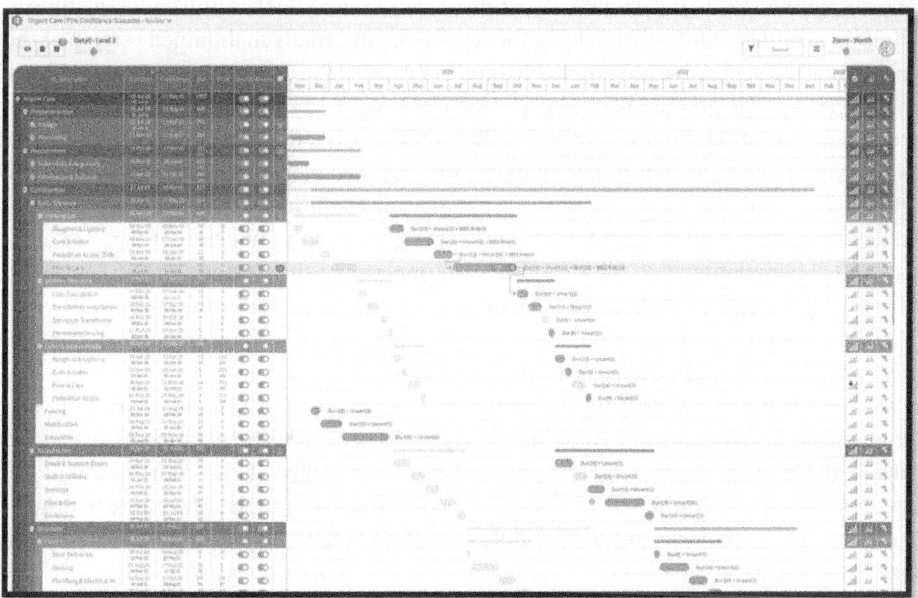

Figure 14.3 Example of a scheduling application.

dates or the duration of the activity. Changing the dates or durations of activities with float to a large enough degree can result in actually changing those activities from not being on the critical path to being on the critical path. Activities that are on the critical path often represent the greatest risk to a project because even a single day of delay to any of those activities will result in a delay to the overall project completion date. This type of scheduling with a focus on the longest sequence of related activities to drive the overall project duration is known as Critical Path Management (CPM).

Very often, the schedule is a contractually required submittal to the project owner, and the contract often specifies how the schedule is to be prepared, such as requiring the identification of the project's critical path. This is so the owner can monitor progress and establish a record of the contractor's plan to perform the work. Such a record can become meaningful when the contractor attempts to submit change orders requesting additional time because the owner can use the contractor's submitted schedules to determine if the claim has impacted anything on the planned critical path, and potentially use such information to deny requests for additional time. It is therefore important for contractors to pay close attention to the information included in the schedules being submitted.

One philosophical shortcoming of the CPM method of scheduling is that the plan is predicated on the optimal sequence of events all happening precisely as planned and leaves no room for deviation without incurring a delay to the overall completion date of the project. This results in a work plan that has a high risk of failure because everything on the critical path will need to go exactly according to plan. Such a planning method puts a high degree of focus on the critical path activities and lessens the focus on the activities that have float. This can result in inefficiencies in other parts of the project, which can be missed profit opportunities. Also, when delays can and eventually do happen to any activity on the critical path, the result is a lot of effort expended in creating various iterations of new schedules and work plans attempting to recover lost days.

When a CPM schedule is a contractual requirement, contractors should try to not be overly optimistic with what is communicated to the owners, rather providing a schedule that utilizes all the time provided in the contract is a sound approach. Managing the work to an internally maintained version of the schedule that includes more aggressive but attainable targets provides the company with a good plan to strive for and challenges crews to do their best. The contractually required schedule provided to the owner can be updated to reflect continual improvements to the overall completion date as activities are completed ahead of the owner's schedule, but in line with the internally managed schedule. Starting a project by providing the owner with an overly optimistic schedule can result in difficulty managing expectations and spending a lot of time explaining why schedule slippages are occurring even if they are not jeopardizing a contractual completion date. This can leave an impression that the contractor does not plan his work well, while conversely, providing schedule updates to the owner that continually indicate improvement to the eventual project completion date conveys a message that the contractor is capable of efficiently executing the work and is exceeding initial expectations.

Scheduling work using a CPM schedule is a very effective tool, but it is just one view of the work plan, and it is from the perspective of time. Additionally, the CPM schedule, or the project's master schedule, is not typically the best tool to manage the day-to-day work because of the effort needed to update a schedule and the complexity inherent in schedule relationships and logic. The effort required to keep that level of detail in the project's master schedule each day is often not worth the effort, and the crews in the field need more basic capabilities to manage their work on a daily basis.

Look-ahead plans or planning boards are a common tool and are used to plan at a more detailed level of *Joe, Jim, and Bob will take the backhoe out on Tuesday and lay 200 feet of water line in Area 5.*

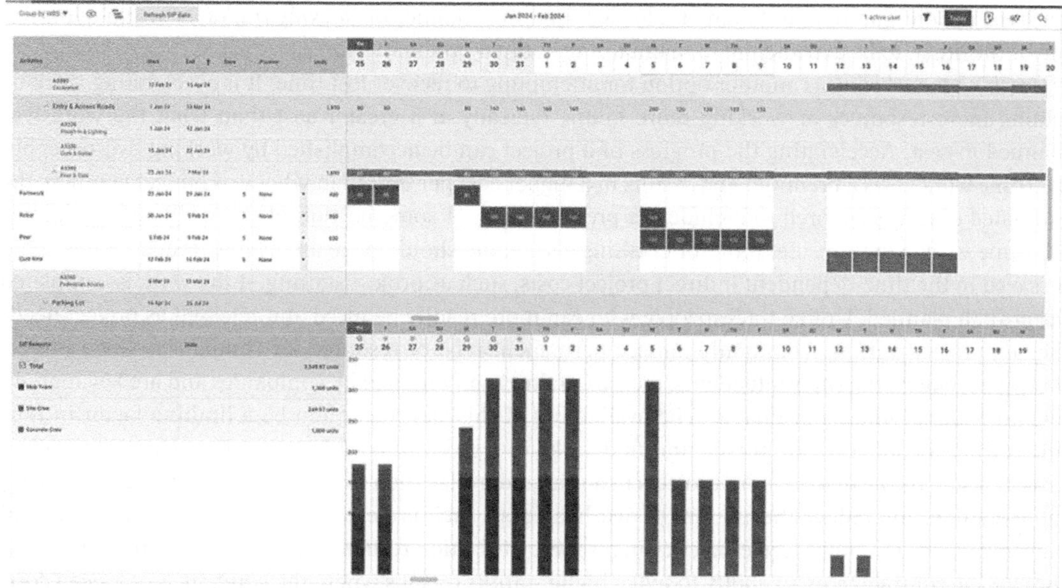

Figure 14.4 Example of a short interval planning application used to plan work on a daily basis.

A whiteboard with dry erase markers or sticky notes in the job trailer is all that is needed to plan at this level, and what should be done periodically is to correlate those look-ahead plans back to the master project schedule to ensure the work being executed in the field conforms with the overall work plan as reflected in the project's master schedule. Computer applications are available for field staff to perform this daily planning effort digitally, and such systems can help automatically identify things like resource overallocation, using the same person or piece of equipment on two different activities at the same time, or breaches of the master schedule where the work plan exceeds the time permitted for those related activities on the CPM schedule (Figure 14.4).

Tracking the progress of various activities in the schedule can alert project managers to work being performed out of accordance with the plan, and the overall impacts of such delays. When identified early enough, such variances can give project teams time to react to and recover from any time lost and maximize the likelihood of putting the project back on track. Corrective actions such as resequencing activities to start other work early or starting various activities at the same time are known as *compressing the schedule* and can be a good way to recover lost time. It does, however, come at the expense of adding productivity risk due to the additional complexity of resequencing work or working activities in parallel, which are optimally planned to start one after the other. For instance, the rough-in of plumbing and electrical work may not have been planned to start until the framing was complete. To compress the schedule, if the project manager directs the rough-in work to start a few days after the framing work has started, the idea would be for the framer to work in a manner that permits the electricians and plumbers to work right behind the carpenters instead of waiting for the carpenters to complete all the framing before plumbers and electricians were permitted to start. Some trades may be able to work effectively in this type of configuration, but many trades will find this manner of *stacking trades,* or working one on top of the other, to be detrimental to maintaining optimal productivity.

Crashing the schedule is another option for attempting to recover lost time. It is accelerating work by adding more resources or working more hours, typically at a greater cost than what the work was planned to cost. Accelerating the progress of a project can be accomplished by working overtime, but overtime work costs a premium and results in a higher cost per unit than what was budgeted unless the estimated cost was factored to include the premium cost of some portion of the work taking place as overtime work. One consideration for crashing a schedule should be to assess how much money *could* be saved in the time-dependent indirect project costs, such as project staffing, if the work is completed faster than planned. Unlike tradespeople who can more readily be hired and released as project needs dictate, project staff are commonly a more permanent type of resource for companies. Good project managers, engineers, superintendents, etc., are trained and retained by companies and are key ingredients to making projects profitable. The availability of this staff can often be a limiting factor in how much work a company can pursue and undertake in a given time. When a company can complete a project faster, they can turn more projects in a year and utilize their project staff to do more work, drive more volume, and reduce the amount of overhead costs each project needs to contribute toward company overhead. This improved efficiency in resource utilization results in a more competitive cost structure for the contractor, or an ability to make higher profits on jobs where the company's overhead costs have already been covered on the previously completed projects. When considering the acceleration costs such as overtime or premium pay, it may require evaluating more than just the increased costs to finish the work sooner. A cost–benefit analysis should be done and be based on a re-estimation of the remaining work and how long to attain overall project completion in order to determine the optimal result for the company.

14.5 Budget

Much like the project schedule, the budget is a reflection of the company's work plan. It is initiated from the estimated cost of the work. For the purposes of managing project costs, it should not include things like markup for overhead and profit; thus, the project budget reflects the estimated cost to perform the work and not the contract price. The difference between cost and price is profit, and less a contribution for company overhead: profit reflects the company's view on how much risk is involved in a project and how much reward they are entitled to earn for taking the risk of contracting to do the project. Including markup for overhead and profit in the budgeted items artificially increases the estimated cost of doing the work. Spreading markup and profit throughout items in the bid is necessary when bidding the work and is dictated by the owners' bidding requirements for furnishing a price proposal. However, cost and price are two different things and should be treated as such.

The budget a contractor uses to manage the project is different from the budget from an owner's perspective. For an owner, their budget is based on the cost the owner pays for the work, which is the contractor's price. It is often reflected in a Schedule of Values or some other method for how the owner plans to pay for the work as the project progresses. Unlike the contractor's budget for managing the cost of the work, the owner's cost, which is the contractor's price, will be inclusive of a proportional spreading of the non-itemized costs estimated and required to complete the work.

As far as the contractor's budget is concerned, there can be other types of costs that typically get spread in a price proposal, such as the contingency costs, which are the costs a contractor has estimated for certain identified risks. These types of contingent costs should actually be carried as line items in the budget to be drawn upon as the risks occur or are mitigated. Hedging productivity rates for risky work is not recommended. As an example, derating production for an excavation item when there is a chance the crews may encounter hard material is not a good way to estimate the cost of the work because the use of a lower-than-optimal production rate would result in too conservative of a budget for that activity if the risk doesn't occur. Risk should be quantified based on the likelihood of it occurring and carried in the budget in its own line item. When such a risk does occur, money can be moved from such a contingency bucket into the underperforming activity where it's needed. This type of budget move is often a managed process and can require management approval, providing an indication to upper management that something on the project is happening, which varies from the plan.

Tracking the cost of the work is commonly performed with a contractor's accounting solution. Accruing costs for invoices that have been received and approved and coding expenses to projects using a breakdown of cost codes is common and well within the capabilities of most commercially available accounting systems, but effectively managing a project budget involves more than simply accounting for the job costs incurred. There are software applications specifically designed to manage budgets for construction projects. Ideally, the solution integrates with the accounting system and uses accounting costs as a source of information, reducing or eliminating the need for double-entry and human error.

Payroll system can provide very detailed and accurate information on the labor costs incurred to date, but are typically only accurate once payroll has been run and the payroll taxes have been processed. Additionally, subcontractor and supplier invoices as entered into the accounting system precisely provide the costs incurred to date, but are highly contingent upon the contractor receiving them and managers approving them prior to entry into the accounting system. For these reasons, there can be significant delays in the timeliness or inaccuracies in the information used when reporting on and managing the budget. These types of considerations need to be understood and incorporated into the process when relying on budget information for making management decisions on the project. Ideally, the budgeting solution includes capabilities to incorporate costs that may have been accrued for work completed, but not yet received. When the actual costs are booked into the accounting system, the budgeting solution can then be reconciled with the accounting system ensuring the accuracy of both systems.

Because the company's accounting system may not produce information as quickly or comprehensively as what is needed in the field, it is important to not let that be a limiting factor and provide project managers with the information they need to manage the work. However, it is also important to ensure that whatever solution used in the field is synchronized with the accounting solution. What cannot happen for an effective project control solution is to have a budget that indicates the project is being well managed and profitable while the company's accounting solution indicates something else. Ultimately, the company's accounting solution will be an accurate reflection of the financial health of the company; the project controls solution needs to be the predictor of financial results and should provide the early warning indicators of plan variances so project managers can understand the impact of issues as they happen in the field and respond in real time based on actionable information (Figure 14.5).

The project budget is derived from the estimated cost to build the work, and it is common for the estimate to have been prepared at a greater level of detail than what is needed for tracking the budget. Retaining too much detail in the project's budget can make it more difficult for the field execution team

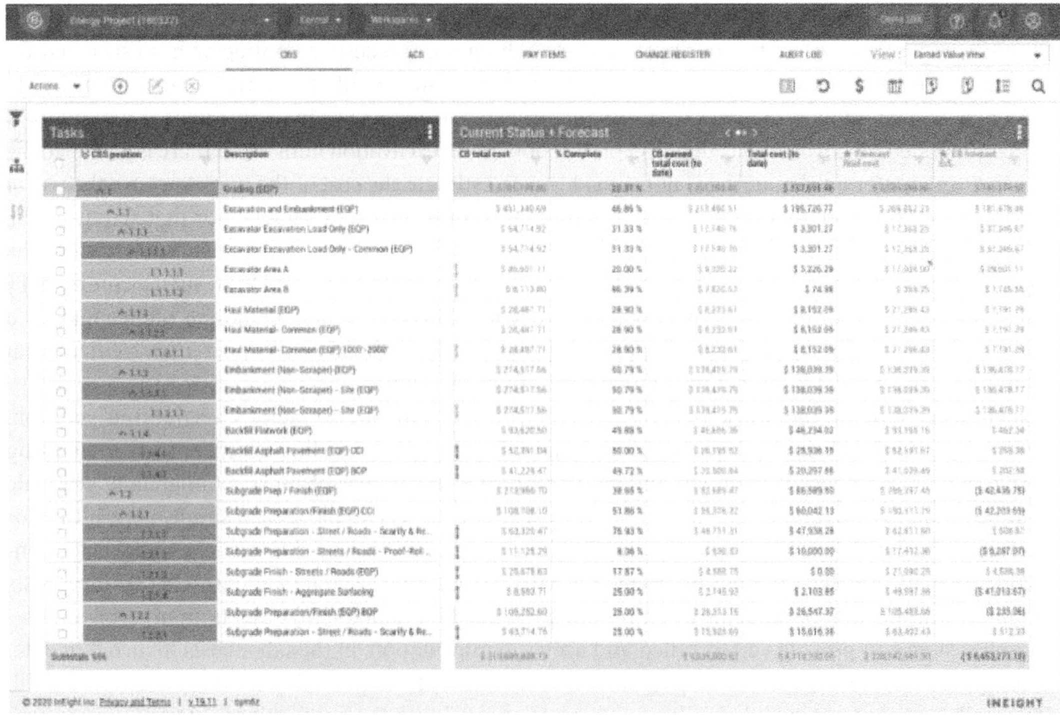

Figure 14.5 Example of a project control budget.

to accurately record costs and progress to the appropriate budget items. As an example, the estimate of a building's foundation system may include a line item for each of individual spread footing. The budget, however, should have a single line item for all spread footings because the additional value in being able to determine one footing costs slightly more than another footing is not likely worth the additional effort required to accurately code the costs and man-hours incurred against each individual footing. In fact, if too much detail is required of the budget, the likelihood of quantities, time, and costs not being recorded against the correct budget codes greatly increases. This is not to say that capturing greater volumes of data will not provide more detail that may be useful as the work progresses and changes to the work plan are needed, but the more data that are required to be captured, the more likely there will be errors in collection of that data. Data collection technologies such as laser scans and lidar can greatly assist in this regard, but when setting up the budget, consideration should be given to the appropriate amount of data capture relative to the effort required to capture it.

A key piece of information that needs to be collected regularly is the measurement of quantities installed and completed to date. Accurate quantities are the single most important means of determining the stage of completion of any particular scope of work and should be used to measure progress against the budget. This is one of the processes where simplifying the budget for field use is appropriate because if the budget is too detailed, the process of measuring the work performed in the field becomes complicated. The level of detail for any activity can be easily adjusted for ease of collecting the data. The greater detail is still of value to the project manager and schedulers. The overall purpose is to be able to make meaningful management decisions based on progress made against the budget.

We should recognize that this is NOT the job of accountants. Ensuring the budget accurately reflects the costs incurred to date is part of a field activity of measuring progress against the budget. Information on actual progress made to date cannot be produced by the accounting department. Accountants have use for the information but collecting it is a field activity as are many other sources of information needing to be provided by the project team, such as seeing that time cards and invoices get coded to the proper budget codes, submitting the billing quantities accurately and on time, processing change order requests as soon as practical, etc. Maintaining the accuracy of costs expended against the budget and the accuracy of all the sources of information contributing to it are critically important responsibilities of the field team. This of course includes ensuring that the budget and adjustment accurately reflects the progress of the project as it's being built.

14.6   Earned Value Management

The requirements for financial reporting are complex, but the calculus for determining profit is not. A shortcoming in using only an accounting-based approach to manage the work is that it does not provide good visibility into the progress of the projects. Understanding how we are doing on a project week to week, even day to day, is critical. It is not much use to field operations to discover how a job came out financially after it is completed. Knowing how well the work is performing while the work is in progress is the essence of good project management.

Effectively managing progress on a project requires an understanding at any point in time of how a project is doing compared to plan with regard to both time and money. Both the budget and the schedule are tools used to manage the progress of a project and reflect the progress against the work plan. However, most of the software applications available are designed to manage either the budget or the schedule separately. Few are capable of managing both effectively. The functions of managing the project schedule and the budget tend to be handled separately. This, while most project decisions impact both the time and cost to complete the work. Time and cost are two sides of the same coin, and if the work plan is to be managed effectively, changing the plan in any way needs to consider the impact on both. Failure to plan the work by assessing the combined impact of any change to both the budget and the schedule means that one may be optimized at the expense of the other.

Earned Value Management (EVM) offers a project management methodology to manage the progress of a project incorporating the interrelated elements of both cost and time. As an example, consider a contractor working to complete 4 miles of laying rail track for a cost of $4 million and has four months to complete the work. If, after three months, the contractor has spent $2 million of the project budget, we can infer they are 50% complete from the perspective of the budget ($2 million spent to date ÷ $4 million total budget = 50%) and 75% complete from the perspective of the schedule (3 months to date ÷ 4 months total = 75%). The problem with either of these assessments is they fail to accurately describe how much work has been completed. If after three months of work, the contractor has only completed laying 1 mile of track, the project would only be 25% physically complete (1 mile built ÷ 4 miles total = 25%). At that current rate of production, the project will cost $8 million and take 12 months to complete. These are traditionally difficult to determine until the planned cost in the budget or the planned time in the schedule has been exceeded.

EVM provides a set of metrics that can be used to easily assess the efficiency of the project's operations, such as Earned Value (EV), Cost Performance Index (CPI), and Schedule Performance Index (SPI). While these terms may not be familiar to all contractors, the concepts they represent are regularly used by most contractors in the management of their work. Contractors assess cost performance by looking at unit costs for work such as *the concrete was budgeted to cost $90/CY and it actually cost $110/CY*; however, they don't necessarily refer to that assessment using the term CPI. They also assess schedule performance by comparing *the planned two weeks for an activity that actually took three weeks to complete*, but they don't refer to that evaluation using the term SPI.

There are plenty of resources available that describe EVM as a project management methodology and it is not the intent of this text to explain the intricacies of EVM. Rather, we offer an explanation of how these principles are applied to manage budget and schedule together and to explain what some of the EVM metrics mean so that readers might better understand systems of project controls – systems that provide information to effectively manage the progress of projects.

14.6.1 Measuring Progress

Accurately determining progress on a project is necessary to provide project managers with the ability to obtain early warning indicators of issues occurring on a job before those issues manifest themselves in conventional manners, such as budget and schedule variances, which may appear on cost reports as exceeding the budgeted amount or the duration of a scheduled activity.

Measuring progress is accomplished by measuring the physical *percentage of completion* of the work being put in place at a particular point in time. For example, a project requiring the installation of 650 cubic yards of foundation wall, and the crews having completed placing 275 cubic yards, then the percentage of completion would be 42.3%.

$$275 \text{ cubic yards, as-built} \div 650 \text{ cubic yards, total} = 42.3\%$$

The physical percentage of completion is then used to determine how much of the budget has been earned or how much of a schedule's activity has been earned. Looking at how much work has been physically completed enables an assessment of operational performance early enough in the project while there is still time to take corrective action for underperforming activities.

How a company chooses to measure progress on a project is a management decision and is not determined by how the contract defines the method of measurement for payment. For example, a unit price contract may impose conditions that benefit the owner from a cash flow perspective but do not effectively measure the actual work completed. A contract may call for pavement in place to be measured by the square yards of surface area installed. This method ignores the impact of varying pavement thicknesses and would not be an accurate measure of the amount of work installed. Thicker pavements require more material, increase the time and effort to place the mix, increase hauling costs, require more compaction, etc. In this case, an appropriate measurement of the progress of the pavement installed would be to measure by the volume of material placed (cubic yards), or even by the weight of material placed (tons). Measuring by weight may be a better method because it can be cross-referenced with external sources, such as delivery tickets from the supplier indicating the quantity of material furnished.

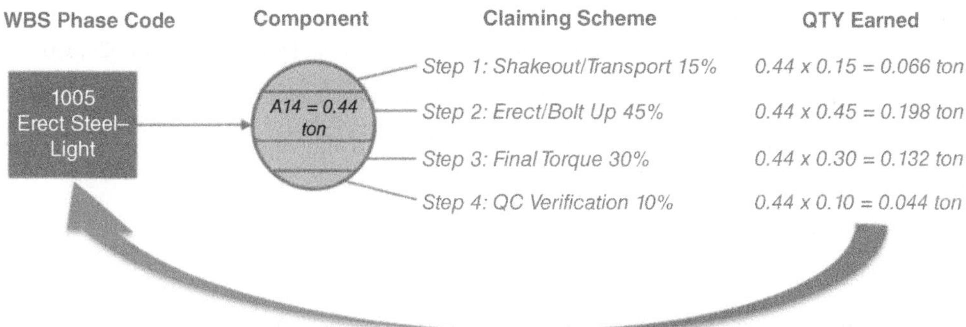

Figure 14.6 Rules of Credit for claiming quantities and calculating percent complete.

For some operations, it can be difficult to measure the progress of work that entails many steps, such as installing structural steel. Steel erection is commonly measured using tons but erecting a single steel beam can entail many steps, such as hoisting the beam into position, securing the beam, lining up and bolting the connections, and torquing the bolts to a specified amount. All these steps would be necessary to complete before claiming an amount of work as being complete, which in this case would be the weight of the single beam.

To simplify the accurate tracking of complex work operations, incremental milestones or *rules of credit* can be used. Rules of credit utilize a list of identified steps by assigning a percentage to each step of the operation's overall completion (Figure 14.6). Progress can be measured by simply checking off each step as they are completed, and the overall percentage of completion for the activity is calculated using the percentages applied for each completed step. It is common to have a small portion of the overall component applied as the last step, such as a testing or inspection step. This prevents the item from being claimed as 100% complete until the work has been accepted as final.

14.6.2 Earned Value and Cost Performance

Once the percentage of completion is established for a budget item, it is easy to determine how much of the budget has been earned. Earnings are the basic building blocks of the EVM approach and enable the accurate assessment of progress relative to the plan. As an example, the steel framing of a small warehouse has a budgeted cost of $100,000. When 36 tons of the 90 tons of steel framing have been erected, the framing is 40% complete.

36 tons, as-built ÷ 90 tons, total = 40%

With the percentage of completion known, we can determine that we have earned $40,000 of the $100,000 budget for the steel framing work.

$100,000 budget × 40% = $40,000 earned value

This earned budget value can now be compared to the actual cost incurred in completing 40% of the work. The actual cost to date of the work performed is obtained from cost reports and, in this example, comprises payroll records for labor costs, invoices received from the material supplier, and rental costs of the crane, and totals $60,000. Because $40,000 of the budget has been earned, we can compare that to the as-built cost to date of $60,000 and determine that we are over budget by an amount of $20,000.

$40,000 earned value − $60,000 actual cost = −$20,000 cost variance

Conversely, if the same project has physically completed 65% of the work, the earned budget amount would be $65,000.

$100,000 budget × 65% = $65,000 earned value

And compared to the as-built cost to date of $60,000, we would be beating the budget by $5,000.

$65,000 earned value − $60,000 actual cost = +$5,000 cost variance

Early understanding of how an item of work is progressing relative to its planned cost gives managers an understanding for how effectively work is being executed and, if necessary, provides the company with time to react. Indicators of underperformance may not always require an immediate response. It is good practice to not rush to judgment on every variance discovered, particularly if it occurs early in the activity. It may be sufficient for the variance to warrant closer monitoring of the operation and ensure there are no specific reasons for the anomaly and simply continue monitoring the operation before determining if intervention is necessary.

For a large, complex project with many different activities being performed at the same time, having such metrics helps quantify the subjective judgments that can be characteristic of progress reports. The metrics can also provide management with a better sense of direction in determining which portions of the project require the most attention.

Determining how much of the budget has been earned and comparing it to how much has been spent is assessing the project's *cost performance*. The more this type of performance assessment can be shared with those individuals actually performing the work, the more likely the project will be to hit its goals. Performance measurements are an ideal source of feedback for crews about operational efficiency. Some organizations have used a motivation process of planning a day's work by letting foremen or crew leaders predict their daily production before attempting the work. The crew then uses the daily planning tools to indicate whether the day's plan will beat the budget or not based on the predicted production output (Figure 14.7). This empowers those responsible for driving the work to participate in planning and goal setting. Most report that this level of involvement has shown significant improvements in productivity

Task ID and Description		Actual		MHRs	MHRs per Qty			Actual	
Task ID	Description	Quantity	UOM	MHRs	Planned	Actual	CB ▾	Cost G/L	MHRs G/L
1005	Erect Steel - Light	4.90	Ton	32	6.202	6.531	30.39	71.5	-1.61
Totals				32				71.5	-1.61

Figure 14.7 Example of a daily plan beating budget on cost but exceeding budgeted man-hours.

among crews. Additional productivity improvements have been reported as crews try to beat their previous best times.

Tracking crew productivity and enabling the workforce to achieve optimal production rates is highly effective, but other elements may also impact cost performance, such as material procurement or cost escalations in labor and material prices. For these reasons, EVM uses a metric called Cost Performance Index, or CPI. CPI is an efficiency measure of spending and indicates that for every dollar spent, the project is earning so many dollars of budget. CPI is calculated as follows:

$$CPI = \frac{Earned\ Value}{Actual\ Cost}$$

Using the preceding example of our steel-framed warehouse, the steel framing work at, 40% complete, would have a CPI of 0.67 and is calculated as follows:

$$CPI = \frac{\$40,000\ \text{earned value}}{\$60,000\ \text{actual cost}} = 0.67$$

This indicates that for every $1.00 spent on steel framing, the company is earning $0.67 of the budget.

14.6.3 Evaluating Cost and Schedule Together

Cost performance is not the only consideration necessary to ensure a project is being controlled effectively. Some projects have plenty of time to do the work, but it is costing too much, or conversely, no issue with the cost of the work but the schedule is challenging. Time and cost are two sides of the same coin but are not mutually exclusive. Making changes to the work plan to improve the schedule commonly means increasing cost, and vice versa, so managers need to assess cost and time together. To that end, a similar metric is used to assess the schedule's effectiveness. Time is measured in days, but to relate the cost and time together, schedule activities are cost-loaded, and the cost of those activities are referred to as *planned values*. Planned values can then be compared to earned values, and this is called the SPI. SPI is calculated as follows:

$$SPI = \frac{Earned\ Value}{Planned\ Value}$$

Assume the steel framing of the warehouse takes 12 months to complete. Six months into the activity, it would be considered 50% complete. The cost of the activity would be the budgeted amount of $100,000 and being halfway through the activity's planned duration would result in 50% of the schedule's planned value being complete, or $50,000.

$$\$100,000\ \text{budget} \times 50\% = \$50,000\ \text{planned value}$$

SPI would then be 0.80 and is calculated as follows:

$$SPI = \frac{\$40,000\ \text{earned value}}{\$50,000\ \text{planned value}} = 0.80$$

Similar to CPI, SPI is used as a measurement of schedule performance, and it shows for every $1.00 planned, the project is earning $0.80 of the scheduled value. Both these indicate if the project is getting value out of the work that is being paid for, in terms of both time and cost.

CPI and SPI of greater than 1.0 indicates an activity is doing better than planned, while CPI or SPI of less than 1.0 indicates the activity is doing worse than planned. CPI and SPI can be calculated for all

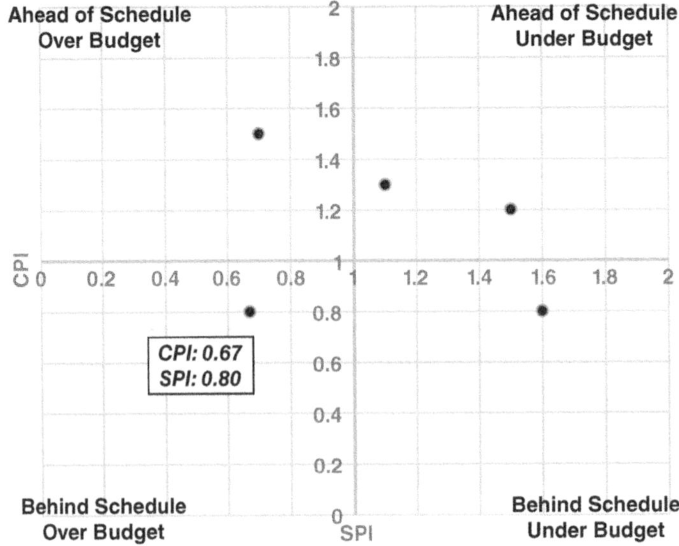

Figure 14.8 Example of a scatter plot of CPI and SPI for budget items/activities.

activities being tracked on the project and plotted on a scatter plot to provide a visual identification of which parts of the project may be under-performing and in most need of attention (Figure 14.8).

Understanding how a project is performing and, in particular, how various activities on a project may be performing provides early indicators of problems that threaten the profitability and timely completion of a project. There are many reasons for cost or schedule variances, such as cost escalations, poor planning, and inappropriate overtime. The first step in addressing these issues is being made aware of them, which in and of itself can be challenging as projects get bigger and more complex.

14.6.4   Forecasting

As activities near completion, especially when the activity is on time and on budget, the risk to that portion of the work is reduced. Therefore, it is important to forecast the effect of an underperforming part of the project as early in their progress as possible in order to assess any impact. A quick and easy way to figure out how much an ongoing activity will cost in total, while it is partially completed, is to divide the budgeted cost of the work by the CPI. The result is the estimated cost at completion. It is a value predicated on nothing more than the average performance of the activity and can represent the total cost of the work if nothing is done and the work continues to progress at the same rate through completion.

The Estimate at Completion (EAC) amount can be calculated as follows:

$$EAC = \frac{Total\ Budget}{CPI}$$

In the example of the steel-framed warehouse and the CPI, the total budget was \$100,000, and with a CPI of 0.67, the EAC would be as follows:

$$EAC = \frac{\$100,000 \text{ budget}}{0.67} = \$149,254$$

While this estimated value is not totally reliable because it is uncertain whether the work will continue at the same rate for the remainder of the project, it is an easy way to estimate the total potential impact of an underperforming item. It is a quick way to evaluate many activities in varying stages of completion. The EVM metric is a tool to manage the work and guide decision-making. Understanding which activities can potentially represent the most budget risk provides management with a way to prioritize their efforts and focus on the most important elements.

A comprehensive use of EVM metrics is to forecast the project's total cost at completion across all activities. If variances are identified and root causes are determined, corrective actions can be undertaken and the estimated cost to complete the work can be predicated on a new work plan and approach. The new estimated cost can be reflected as the new forecasted cost to complete the work and can be monitored using those new assumptions.

When forecasting costs to complete, average performance is a great starting point, because the organization has already proven they can accomplish that level of productivity. However, it is also important to consider the various factors that decide if using average performance would be appropriate. Is the most difficult portion of the project behind us or do we expect that the more difficult work is yet to come? Are we past the initial learning curve where average performance would be appropriate to use or should we continue using our planned production rates until we can attain a steady state of production? For activities that are early and have yet to achieve an optimal production rate, use budgeted production rates to forecast the costs to completion until the activity is at least 10% complete, or some percentage that makes sense for the work being considered.

When forecasting costs to complete, judgment needs to prevail, and the forecast of the overall cost to complete the work should be based on an analysis of the work in progress and already performed. Apply the as-built productivity rates where appropriate and incorporate the newly estimated costs and rates where the work plan has been modified. If the total forecasted cost to complete the work seems unreasonable, it is an indication that further analysis should be performed in an attempt to discover any root causes or issues threatening the project (Figure 14.9).

14.6.5 Reporting

The output of an EVM system is the reports, which may be used for a variety of purposes, but mainly they are to provide management with quantitative data that can be used in supporting project-level decisions. Some federal projects require EVM reports as submittals, but the real value of EVM is in providing management with quantifiable metrics and better insights to manage projects. Metrics such as CPI and SPI assess the performance of particular activities, and the aggregation of all activities is indicative of how well the overall project is progressing. S-curves are a commonly used visual tool to communicate the consolidated information for the entire project. Cumulative budget, earned value, and actual costs incurred

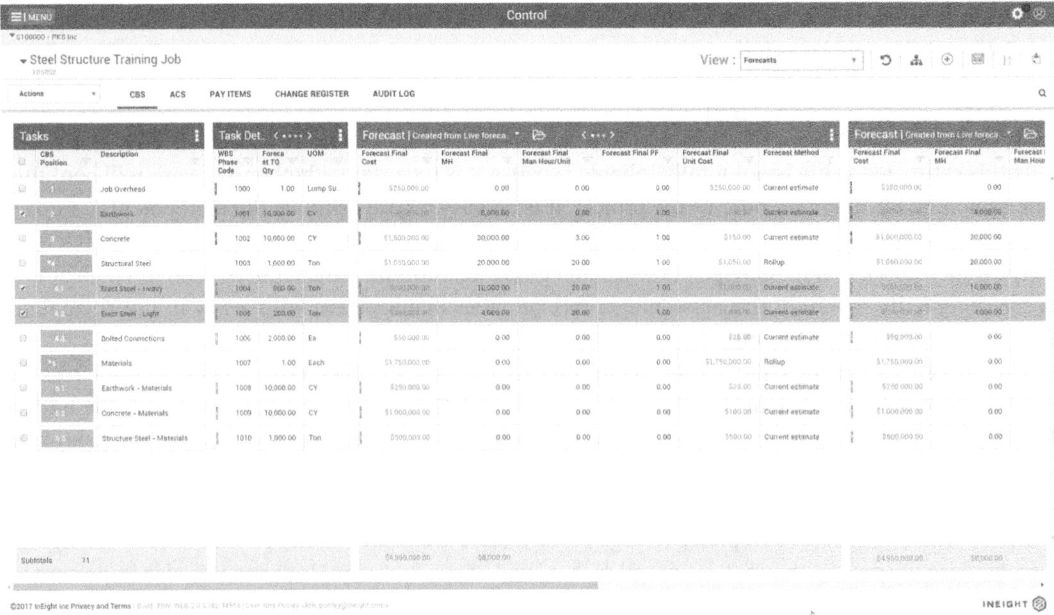

Figure 14.9 Budget with forecasted cost to complete.

can be plotted over time, and planned values, represented by dashed lines, can highlight variances and show the trajectory of the expected S-curve based on the project plan. Dashboards can bring together all the necessary information from a variety of different sources and provide managers with a consolidated view of the overall project performance (Figure 14.10). This provides the ability to drill down into the various aspects of the project to better understand the possible reasons for identified variances.

In addition to effectively monitoring the progress of projects, EVM provides another source of information for the financial management function of the company. Preparing a WIP schedule relies on a re-estimation of the remaining work to be completed for all work in progress. An inaccurate estimation of the remaining cost of the work can have a dramatic impact on the accuracy of the calculations and results of the WIP and in turn the company's resulting financial statements.

When preparing the WIP, most companies don't actually re-estimate the cost of completing work, often because the estimating department is busy getting future work. One of the most difficult parts of estimating is attempting to accurately predict the cost of future work, but once the work has begun, many of the variables that make estimating difficult have been removed. The materials have been purchased or at least the prices should be well known; the crew has been determined, and if work has begun, there is a known and achievable production rate that management can use as the basis for predicting future costs. The estimated cost to complete the work is a by-product of an effectively implemented EVM process and represents a more accurate re-estimation of the work because it incorporates as-built costs to date and forecasted costs to complete based on up-to-date costs and production rates. At a minimum, re-estimating costs on the project should be focused on the areas of the project that have yet to be

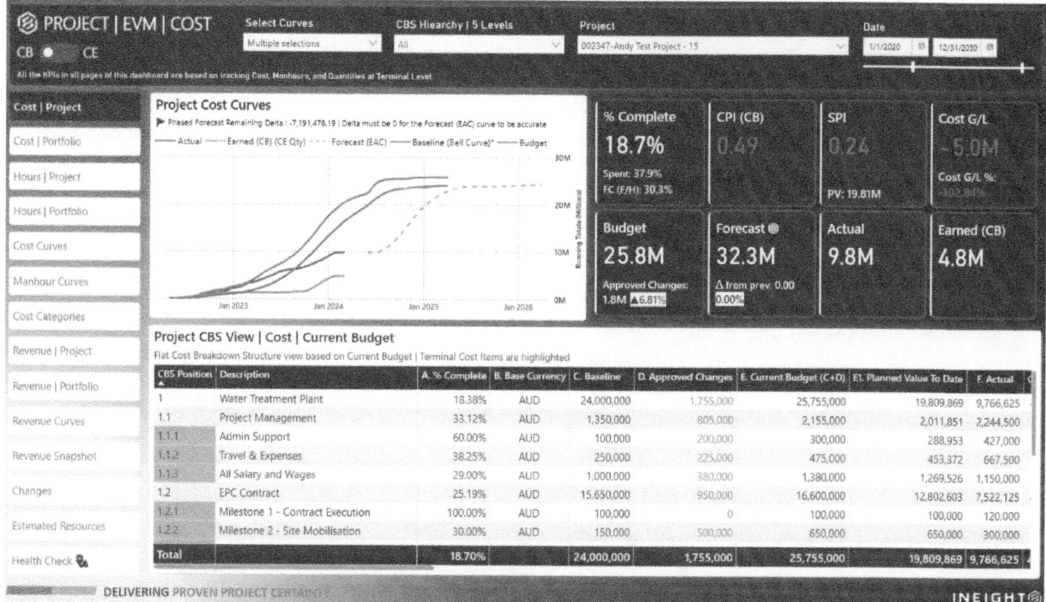

Figure 14.10   EVM dashboard displaying an S-curve.

completed and have been identified as having the most significant impact on the project's total cost, which is where the attention and involvement of management will be most impactful.

14.7   Conclusion

Project controls are a project management methodology providing companies with a consistent system and tools for measuring progress on projects. When effectively implemented, these tools reduce the subjectivity inherent in answering the question, *how is the project going?* Companies that are growing will need to rely on a greater number of individuals to manage the work, and having a system in place to quantifiably measure performance reduces individual subjectivity and provides meaningful ways of assessing performance and operational efficiency.

As discussed earlier in this book, bad projects almost never get better on their own and often do not improve even when the organization is doing everything it can to turn things around. The best defense for a losing project is to not take it in the first place, but if you do have a poor-performing project, learning about problems early enough to do something about them provides the best opportunity for recovery. Contractors need a system in place where performance is tracked and forecasted consistently across multiple projects. Management can then monitor performance indicators, raising alerts when thresholds are exceeded.

Chapter Review Questions

Chapter 1

Review Questions

1 Which of the following industries has the highest rate of failure?
   A Automotive
   B Oil and Petroleum
   C Banking
   D Construction

2 Which of the following is a true statement?
   A Many construction projects are not very risky
   B The reasons for a construction business failure often occur years before the failure itself
   C Construction firms that have been in business for a long time are less likely to fail
   D Construction business failures can be avoided by completing all projects on time

3 The primary goal of a construction companies' leadership should be to earn a profit
   A True
   B False

4 Which of the following statements about construction company growth is most accurate?
   A Bidding work more aggressively increases a company's profit margin
   B Construction companies only grow when the market is shrinking
   C Growing in a steady market can only happen if your competition obtains less work
   D Bidding more work can only be accomplished when the market is growing

5 According to the research referenced in the chapter, which of the following would be a more common reason for a business failure?

A Lack of effective managerial control

B Failing to bid enough work

C Executing contracts with unfavorable terms and conditions

D Performing risky work

Critical Thinking and Discussion Questions

1 Describe some of the more recent evolutions of the construction industry and explain what effect has been on contractors' ability to be profitable.

2 As referred to in the chapter, give three examples of a company *business practice*.

3 Explain what is meant by the statement "the differentiator for profitable contractors is the quality of their of business practices and not the quality of their work."

4 Describe the three main functional areas of a construction business and explain why it's difficult for a firm to be effective in all three.

5 Explain why it is so challenging for construction companies to be a profitable business today.

Chapter 2

Review Questions

1 The lack of a formal organizational structure is a weakness among construction companies.

A True

B False

2 Which of the following statements is a best practice when hiring a new manager?

A Hire an individual who has similar skills as the company owner because they are more likely to work better together

B Hire an individual who has different but complementary skills from the company owner because they will increase the collective skillset within the company

C The similarity of skills of the individual being hired is unimportant because all major decisions within the company should be made by the company owner anyway

D All of the above

3 Managing the business side of a construction company is the same for all but which of the following?

A Road builder

B Electrical subcontractor

C General building contractor

D All of the above

4 Which of the following would *not* be considered an overhead cost for a typical construction company?
   A Accounting
   B Estimating
   C Project manager
   D Marketing

5 Which of the following roles support producing the work in the field?
   A Receptionist
   B Accountant
   C Estimator
   D All of the above

Critical Thinking and Discussion Questions

1 Explain how the owner of a construction company can work with managers to maximize the quality of important decisions that need to be made within the company.
2 Describe how to tell when a small company becomes a medium company and when a medium company becomes a large company.
3 Define the term incidental-growth pattern.
4 Describe how flexible overheads can help protect a construction company in times of market volatility.
5 What is the primary objective of a construction business?

Chapter 3

Review Questions

1 A weather event occurs, delaying the progress of a construction project, resulting in the contractor incurring additional costs to complete the work on time. As defined in the chapter, this would be an example of which of the following types of risk?
   A Cost of risk
   B Insured risk
   C Uninsured risk
   D All of the above

2 Which of the following situations can go undetected for years but lead to the failure of a construction business?
   A A sharp increase in material costs
   B Suddenly running out of working capital
   C A subcontractor walking off the job
   D Losing too much work to aggressive competitive bidding

3 Which of the following are common responses to decreases in working capital?
   A Increasing the amount of money borrowed on a Line of Credit
   B Paying accounts payable (subs and suppliers) more slowly
   C Bidding more aggressively to increase sales volume
   D All of the above

4 According to the research, which of the following rates is a reasonable rate of growth for a company aiming to minimize business risk?
   A 5%
   B 15%
   C 25%
   D 50%

5 Healthy organizational growth is best accomplished by first capturing more work than expanding the managerial resources needed to perform the work.
   A True
   B False

Critical Thinking and Discussion Questions

1 Define what is meant by "hidden risks" and describe how impactful it can be to a construction firm.
2 Explain what it means for a construction company to be undercapitalized.
3 Compare and contrast the sales transaction of a construction firm with that of a manufacturing firm.
4 Define the term "Operational Performance."
5 Explain why a company is more likely to fail during a market recovery.

Chapter 4

Review Questions

1 Which of the following is the most significant contributor to company failure?
   A Decrease in project size
   B Change in type of work
   C Retention of key personnel
   D Managerial maturity

2 For a company that specializes in a certain type of work, which of the following are true of projects that are larger in size than the size of a project the company typically performs?
   A Profit margins are often lower
   B Payments can be slower and retainage can be higher
   C More resources are needed to handle additional requirements
   D All of the above

3 For a company considering bidding on a new project, which of the following would be considered "outside" of that company's typical geographic area?

A A project in the next county

B A project in the next state

C A project more than 200 miles away

D A project outside of the area the company normally works in

4 What is the *primary* objective of the management of a construction business?

A Process invoices and pay the bills

B Support the field side of the business

C Prepare estimates for upcoming work

D Ensure consensus and delegate responsibility to middle management

5 Which of the following functional areas of a construction business are regularly overlooked/underrated by the key people in the organization?

A Marketing

B Estimating

C Operations

D Administration

Critical Thinking and Discussion Questions

1 Describe the three main functions of a construction company and in what roles the company is responsible for seeing these functions are accomplished.

2 Explain the challenges posed to a company taking on a project that is significantly larger than the work it typically performs.

3 Describe some of the ways a company working in a new geographic area can lead to business failure and how can a company mitigate the risk of failure.

4 Describe how public work and private work differ using an example and explaining how a contractor may be ill-equipped to do one vs the other.

5 Describe what is meant by the term *managerial maturity*.

Chapter 5

Review Questions

1 What size project would be considered a large project?

A $5 million

B $50 million

C $500 million

D 50% larger than the size of a project the company typically performs

2 Which of the following are reasons contractors bid on projects larger than size of work they typically perform?

   A Running out of work

   B Declining market

   C Increased competition

   D All of the above

3 Which of the following is a true statement about rapid growth?

   A Growing too rapidly is always fatal

   B Growing too rapidly is always risky

   C Growing too rapidly is the only way to keep pace with a growing market

   D All of the above are true

4 Which of the following are valid concerns when taking on larger projects of a similar type of work?

   A Costs per unit are always cheaper for larger projects

   B Working with new subcontractors and suppliers may be required

   C Inspectors typically cannot keep up with the work required of a larger project

   D Owners and owners' representatives are less involved in larger projects

5 Which of the following overhead costs would be considered fixed and thus would be difficult costs to reduce in a declining market?

   A The services of a temporary agency

   B Full-time office employee

   C Short-term leases on equipment

   D Rented office space

Critical Thinking and Discussion Questions

1 What are some of the reasons contractors will attempt work that is larger than the type of size of jobs they typically perform?

2 Using information from the chapter and case studies, provide some examples of the ways larger projects can cause differences in how the work is built compared to smaller projects of the same type.

3 Describe some of the consequences to a contractor failing to perform on a project that is larger than the company may be used to.

4 What are some of the things a contractor should strive to learn about new clients they will be working for, specifically if the project is larger than normal?

5 Describe ways a contractor can contend with a declining market or a reduction in the amount of projects available to bid on.

Chapter 6

Review Questions

1 A contractor's *Normal Business Area* can be defined as which of the following?
  A 10 mile radius from the companies' home office
  B Within a three-county area from the companies' home office
  C The geographic areas a contractor is successful in bidding and winning profitable work
  D On the same side of the river as the companies' home office

2 Which of the following are common reasons for contractors to expand their geographic area?
  A Attempting to obtain a new large project
  B Experiencing an expanding local market
  C Expanding into new areas is a proven risk mitigation strategy
  D All of the above

3 Which of the following are issues commonly associated with expansion into a new area?
  A Availability and qualifications of local labor
  B No experience with local subs and suppliers
  C Lack of understanding of local political climate
  D All of the above

4 Identify an effective strategy contractors can use to mitigate the risk of expanding into a new area.
  A Obtain a large project to ensure there is sufficient revenue to justify the risk
  B Obtain a small project to limit exposure to potential loss
  C Only hire local employees to manage and run any new projects
  D All of the above

5 A contingency plan includes which of the following?
  A A predefined plan identifying when an expansion can be deemed unsuccessful
  B Identification of financial goals and dates for specific accomplishments
  C Specific steps to be implemented upon withdrawal from an area
  D All of the above

Critical Thinking and Discussion Questions

1 Describe some of the reasons a contractor would take on work in a new geographic area.
2 Using examples from the case study, identify and describe some of the more common risks associated with taking work in a new geographic area.

3 When a contractor takes on work in a new geographic area, describe some of the possible impacts on the projects that may be underway in the contractors' normal work area.

4 What are some of the strategies a company can use to expand its geographic area?

5 Explain what a contingency plan is and why it would be important for a contractor to have a contingency plan when expanding into a new territory.

Chapter 7

Review Questions

1 What are the most important questions to ask about pursuing a new type of work?
- **A** Can the company prepare an accurate proposal?
- **B** Can the company complete the work on time?
- **C** Can the company produce the work for a profit?
- **D** Is the new type of work similar enough to the company's normal work?

2 What determines a contractors' specialization?
- **A** The largest of a certain type of project a contractor has completed
- **B** The type of work a contractor ordinarily completes successfully and for a profit
- **C** The companies' NAICS code
- **D** All of the above

3 In the case study of the sewage contractor, what was the primary reason for their failure?
- **A** Installing gravity sewer line was outside of the company's expertise
- **B** Installing ductile iron force mains was outside of the company's expertise
- **C** Installing sewerage treatment plants was outside of the company's expertise
- **D** None of the above, the company was able to profitably complete all types of projects

4 In the case study of the mechanical contractor, what was the primary reason for their failure?
- **A** Selling off the HVAC division
- **B** Selling off the mechanical division
- **C** Selling off the service division
- **D** Co-locating the estimators and contract managers in an office park

5 What is an effective strategy for maintaining a company's profit margin in a declining market?
- **A** Pursue larger projects to increase sales volume
- **B** Expand operations to service a large geographic area
- **C** Diversify the mix of services the company offers by performing different types of work that are similar to the company's normal work
- **D** Allow sales volume to decrease and reduce overhead until the market returns to pre-declining levels

Critical Thinking and Discussion Questions

1 Explain how a contractor lacking experience in a certain type of work would have difficulty in expanding into that type of work.
2 Describe how a contractor can be specialized in a type of work without necessarily recognizing they are indeed specialized.
3 Describe the implications of a contractor attempting to switch from either a union shop to a merit shop and vice versa.
4 Explain why it is critical for a contractor to understand what type of work they are good at before expanding into a new type of work.
5 Describe how a market slowdown can be an opportunity for a contractor.

Chapter 8

Review Questions

1 Which of the following functional areas in a construction company are most susceptible to a negative impact due to a change in key personnel?
   A Getting the work
   B Doing the work
   C Accounting for the work
   D All of the above

2 Which of the following are true of the company values that contribute to the long-term success of a business?
   A Values are instilled in the organization by the founder
   B Values are voted on by the board of directors
   C Values never survive the succession of the founder
   D Values are not necessary for a company to be successful

3 A company founder who is no longer actively involved in the company no longer has any influence on the company.
   A True
   B False

4 When a company loses a key person, what should they do to reduce risk upon installing new management?
   A Bid more work to increase sales volume
   B Limit growth or decrease sales volume
   C Subcontract any work that is typically self-performed
   D Promote all new management positions from within the company

5 As a company grows and needs to introduce more management, key personnel may no longer work with long-time employees, which can lead to employees feeling less appreciated and dissatisfied. This is known as

A Micromanagement
B Collaboration
C Dilution
D Hero Worship

Critical Thinking and Discussion Questions

1 Explain how the loss of a key person can bring about the failure of a construction business.
2 Describe why it is difficult for a successful company, founded by multiple partners, to survive a breakup of the partnership.
3 Explain why it is so difficult to replace a key person in a company.
4 Describe how replacing key personnel can be thought of as essentially creating a new company.
5 Describe one potential strategy to mitigate the risk to the company of losing a key person.

Chapter 9

Review Questions

1 Which of the following skills are developed and responsible for a company evolving from a small startup to a large construction enterprise?

A Business skills
B Building skills
C People skills
D Computer skills

2 Which of the following issues are typical of a company attempting to do more work than their team is capable of effectively managing?

A A decrease in customer complaints
B An increase in absenteeism and turnover
C A reduction in job site accidents
D An increase in field productivity

3 True delegation of authority happens when

A A new manager is promoted from within
B A new manager is hired externally
C A new manager is permitted to make a mistake and live with the outcome
D A current manager is replaced by artificial intelligence

4 Which of the following can be lost when a key person leaves a company?

A An intuition that identifies unique project risks

B The ability to handle a particular type of customer

C Effective resolution of personnel issues

D All of the above

5 What is the most frequent challenge faced by companies replacing top management?

A Lack of a formal succession plan

B Existing management team resents the new personnel

C New personnel are overqualified for the position

D New personnel lack confidence in the existing personnel

Critical Thinking and Discussion Questions

1 Explain how the management of a small start-up company is different than the management of a growing construction business.

2 Describe some of the common signs that occur when a company is struggling with managerial maturity.

3 Explain how changes in top management can put the entire company at risk.

4 What is delegation of authority and describe how it can be effectively accomplished?

5 When training new leadership, explain why the *let them learn it the way I did* approach does not result in effective succession planning.

Chapter 10

Review Questions

1 Which of the following capabilities are most important for a contractor to be successful?

A Obtaining the work

B Making a profit

C Building the work

D Hiring a CPA

2 Which of the following is NOT an annual financial statement prepared by an independent third party?

A Balance sheet

B Income statement

C Statement of cash flow

D Job cost report

3 An approved change order would be reflected in which of the following Work in Progress categories?

   A Contract price
   B Direct cost to date
   C Estimated cost to complete
   D Amount billed to date

4 Which of the following Work in Progress categories are the most difficult to report but are most critical for accurate reporting?

   A Contract price
   B Direct cost to date
   C Estimated cost to complete
   D Amount billed to date

5 The cost of a subcontractor's charges for necessary additional work would be reflected in which of the following Work in Progress categories?

   A Contract price
   B Direct cost to date
   C Estimated cost to complete
   D Amount billed to date

Critical Thinking and Discussion Questions

1 Describe the differences between internal and external financial statements and explain how they are used.
2 Explain why it is critical to have input from the people engaged in the work when preparing a company's financial statements.
3 Explain how GAAP rules make it challenging for a construction company to produce highly accurate financial statements.
4 Compare and contrast the accounting for work in progress of a construction firm with the accounting for work in progress of a more conventional business interest such as a manufacturing firm.
5 Explain how declining profit margins can be difficult to detect in a growing construction company.

Chapter 11

Review Questions

1 Market downturns typically occur about once every

   A Year
   B 10 years
   C 50 years
   D There has not been a market downturn since World War II

2 Which of the following is typically a company's largest overhead expense?

   A Employee wages

   B The rent on the company's home office

   C A company's general liability insurance policy premium

   D Season tickets to the local football team

3 Which of the following would be an example of flexible overhead?

   A A long-term lease on the companies' office space

   B Purchasing a company vehicle for the project manager

   C Buying a new laptop computer for the estimator

   D Hiring a temporary employee to help with filing

4 Which of the following strategies would be the most effective for a company utilizing a piece of equipment for about 20% of the year?

   A Owning

   B Leasing

   C Renting

   D All of the above

5 When a market downturn results in a company not enough of the work it typically performs, which of the following options should it consider

   A Pursue more of a different type of work

   B Pursue more work in a wider geographic area

   C Bid the little work that is available more aggressively

   D Reduce overhead and the amount of work required to support the overhead

Critical Thinking and Discussion Questions

1 Describe the problem with attempting to maintain a certain level of sales volume in a declining market.

2 Explain what it means to downsize a company.

3 Describe how flexible overhead can help a company remain profitable during a market downturn.

4 Provide an example of how flexible overhead can be practically implemented for office personnel.

5 Explain the issues that happen with a growth-driven business model in a cyclical market.

Chapter 12

Review Questions

1 What has to happen when a contractor grows at a rate faster than the market they are in?

   A The contractors' profit margins increase

   B The contractor must raise his prices

   C The contractor must sell some equipment

   D Other contractors have to take less work

2 Which of the following is a false statement that represents a closely held industry believe?

   A Unprofitable projects happen and cannot be avoided

   B A growth rate that keeps pace with the market is sustainable

   C Flexible overhead reduces the need to maintain a certain volume

   D The construction industry is cyclical

3 What is the best way to avoid losing money on the type of project a company has little experience with?

   A Hire a knowledgeable superintendent

   B Increase the bid price to build in a cushion

   C Have the company owner personally oversee the project

   D Don't do the project

4 Which of the following are challenges associated with doing unprofitable work?

   A It is very difficult to turn an unprofitable job into one that is profitable

   B It is more expensive to manage unprofitable jobs

   C It is not possible to minimize costs already incurred

   D All of the above

5 What is the main element in determining project risk?

   A Project size

   B Contractors' experience

   C Location of project

   D Bonding capacity

Critical Thinking and Discussion Questions

1 Describe the implications of a contractor growing faster than the growth rate of the market they are in.

2 Explain what is meant by the statement, *there are no bad projects, just bad matches between contractors and projects.*

3 What is meant by the phrase *experience is accumulated institutionally but is captured individually*?

4 Explain how the risk involved with doing a *large project* is *relative* to the company pursuing the project.

5 Describe how unusual features, such as curved walls, can increase the risk of a project.

Chapter 13

Review Questions

1 What is the single most important factor in selecting a project a contractor is most likely to successfully complete?

   A Amount of experience with similar work

   B The financial capacity to bond the project

   C Relationships with local subs and suppliers

   D A knowledgeable superintendent

2 Which of the following factors should be considered when determining if a prospective project is a good match for a firm based upon their previous project experience?
  A Size of project
  B Geographic location of the project
  C Type of project
  D All of the above

3 Average-size projects can best be described as which of the following?
  A Small but highly profitable jobs
  B Nuisance work
  C Jobs that tend to be easy, relatively undemanding, and mostly perform as expected
  D Jobs that support growth and require the attention of senior management

4 In which of the following Project Selection Program categories would you score the potential risk of an owner who may not possess the ability to adequately fund the project?
  A Project fit
  B Firms' experience
  C Project complexity
  D Financial and cash flow impacts

5 In which of the following Project Selection Program categories would you score a contractor's experience with the projects' designer/architect/engineer?
  A Project fit
  B Firms' experience
  C Project complexity
  D Financial and cash flow impacts

6 In which of the following Project Selection Program categories would you score existing site condition issues, such as a high-ground water table?
  A Project fit
  B Firms' experience
  C Project complexity
  D Financial and cash flow impacts

Critical Thinking and Discussion Questions

1 Describe how the Project Selection Program can be an effective tool for contractors screening potential projects.
2 Explain how a single project can be a great match for one company and a complete mismatch for another.
3 Explain how the size of a project should be a consideration in the appropriate selection of a project.
4 Provide an example of a type of project that may be a good project for one contractor but a bad match for another.
5 Describe how a contractor working in a different geographic location can be a risk factor for selecting a project.

Chapter 14

Review Questions

1   Which of the following roles are typically responsible for ensuring a project is completed on time and within budget?
    A Project Engineer
    B Project Manager
    C Chief Financial Officer (CFO)
    D Chief Executive Officer (CEO)

2   Over the course of an activity, when would the maximum productivity output likely be obtained?
    A On start up
    B Throughout the learning curve
    C After the learning curve but before wrap-up
    D At the conclusion of the activity during wrap-up

3   Which of the following is commonly produced using the Critical Path Management (CPM) methodology?
    A Gantt chart
    B 3-week look-ahead plan
    C Short internal plan
    D Daily plan

4   An activity's physical percent complete is most accurately calculated by which of the following methods?
    A Dividing an activity's cost incurred to date by the total budgeted cost
    B Dividing an activity's quantities completed to date by the total quantity
    C Dividing an activity's days expended by the total activity duration
    D Performing a complete re-estimation of the entire project

5   The budget item has a total cost of $80,000. Calculate the Cost Performance Index for the budget item if the item is 50% complete and the As-Built costs incurred to date are $38,000.
    A 0.50
    B 0.95
    C 1.052
    D 2.105

Critical Thinking and Discussion Questions

1 Explain what is meant by the term *Project Controls*.
2 Compare and contrast the differences between a construction company as a business interest and a more conventional business such as manufacturing.

3 Define benchmarking and explain the differences between benchmarking on costs per unit and benchmarking on productivity rates.

4 Explain why it is important to not let a lack of accounting information be a limiting factor in managing the budget.

5 Explain how the earned value management tools can help contractors with the preparation of the company's work-in-progress (WIP) statement.

Answer Key for Chapter Review Questions

Chapter 1

1 (d);   2 (b);   3 (a);   4 (c);   5 (a)

Chapter 2

1 (a);   2 (b);   3 (d);   4 (c);   5 (d)

Chapter 3

1 (c);   2 (b);   3 (d);   4 (b);   5 (b)

Chapter 4

1 (b);   2 (d);   3 (d);   4 (b);   5 (d)

Chapter 5

1 (d);   2 (d);   3 (b);   4 (b);   5 (b)

Chapter 6

1 (c);   2 (a);   3 (d);   4 (b);   5 (d)

The Business of Construction Contracting: Schleifer's Guide to Financial Success, First Edition. Thomas C. Schleifer and Aaron B. Cohen.
© 2025 John Wiley & Sons, Inc. Published 2025 by John Wiley & Sons, Inc.

Chapter 7

1 (c);   2 (b);   3 (b);   4 (c);   5 (d)

Chapter 8

1 (d);   2 (a);   3 (b);   4 (b);   5 (c)

Chapter 9

1 (a);   2 (b);   3 (c);   4 (d);   5 (a)

Chapter 10

1 (b);   2 (d);   3 (a);   4 (c);   5 (b)

Chapter 11

1 (b);   2 (a);   3 (d);   4 (c);   5 (d)

Chapter 12

1 (d);   2 (a);   3 (d);   4 (d);   5 (b)

Chapter 13

1 (a);   2 (d);   3 (c);   4 (d);   5 (b);   6 (c)

Chapter 14

1 (b);   2 (c);   3 (a);   4 (b);   5 (c)

Index

The Business of Construction Contracting: Schleifer's Guide to Financial Success, First Edition. Thomas C. Schleifer and
Aaron B. Cohen.
© 2025 John Wiley & Sons, Inc. Published 2025 by John Wiley & Sons, Inc.